T0277013

Pineros

Brinda Sarathy

Pineros
Latino Labour and the Changing Face of Forestry in the Pacific Northwest

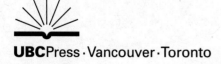

UBCPress · Vancouver · Toronto

21 20 19 18 17 16 15 14 13 12   5 4 3 2 1

Printed in Canada on FSC-certified ancient-forest-free paper
(100 percent post-consumer recycled) that is processed chlorine- and acid-free.

Library and Archives Canada Cataloguing in Publication

Sarathy, Brinda
Pineros : Latino labour and the changing face of forestry in the Pacific Northwest /
Brinda Sarathy.

Includes bibliographical references and index.
Issued also in electronic format.
ISBN 978-0-7748-2113-1

1. Foreign workers, Latin American – Northwest, Pacific. 2. Forests and forestry –
Employees. 3. Foreign workers – Legal status, laws, etc. – United States. 4. Forests
and forestry – Employees – Legal status, laws, etc. – United States. 5. Wages –
Foreign workers – United States. 6. Forests and forestry – Employees – Health
and hygiene – United States. I. Title.

HD8081.H7S27 2012   331.6'368073   C2011-906191-0

Canada

UBC Press gratefully acknowledges the financial support for our publishing program
of the Government of Canada (through the Canada Book Fund), the Canada Council
for the Arts, and the British Columbia Arts Council.

This book has been published with the help of a grant from the Canadian Federation
for the Humanities and Social Sciences, through the Aid to Scholarly Publications
Program, using funds provided by the Social Sciences and Humanities Research
Council of Canada.

UBC Press
The University of British Columbia
2029 West Mall
Vancouver, BC V6T 1Z2
www.ubcpress.ca

To Karthick, Omji, and Millan

Contents

Illustrations

Tables

Preface

In the summer of 2000, before starting my first year as a graduate student at the University of California, Berkeley, I attended Forestry Camp in Plumas County in northern California. For six weeks, I delighted in the smells of Douglas and white fir, lolled in the shadows of enormous sugar pines, measured the height and girth of trees, acquired skills in identifying native plants, and learned about forest management practices. I also saw first-hand the effects of fire on varying forest types and, while attempting to assess forest inventories, experienced the corporal challenges of walking through dense thickets of second- and third-growth stands.

Our class discussions and field trips focused largely on the importance of managing forests for ecosystem health – a departure from decades of managing federal lands for timber – and touched on a variety of techniques ranging from thinning small-diameter trees to introducing controlled burns. At camp, we met with Forest Service employees, loggers, and community members from the nearby town of Quincy, all of whom shared their perspectives on forest management and how it had evolved over time. By the end of summer, I felt confident that I had at least some grasp of the techniques used to maintain healthy forests. Thinning dense forest stands and reducing fuel loads were clearly a part of this process, yet we never witnessed the actual practice of thinning or learned who might be doing this work and how.

Two years later, I chanced upon a collection of oral histories entitled *Voices from the Woods,* published by the Jefferson Center for Education and Research, a small non-profit run out of Wolf Creek, Oregon. *Voices* chronicled the experiences of people, many of whom were non-white and immigrants, in their relationships to forest work – collecting mushrooms, harvesting floral greens, or planting trees and thinning dense underbrush – in Washington, Oregon, and northern California. What astonished me most about these stories was that they represented aspects of forest work and groups of forest workers with whom I was completely unfamiliar. Until that point, the only narratives about forest labour that I had encountered involved the experiences of white, native-born loggers. Were the stories in *Voices in the Woods* an exception?

It has now been nearly a decade since I first heard about non-white immigrants labouring in the woods. Through the course of my research, I have come to learn that Latino immigrants, in particular, are anything but an exception in the forest labour force. Yet the story of how these workers came to "manage" (in the sense of performing manual labour) federal land in much of the Pacific Northwest is largely unknown to the general public – including those who live in the vicinity of these areas. How, then, does one tell a contemporary history that excavates the experiences of workers, many of whom are undocumented and live in the "shadows" of society?[1] Where can one find archival records on such a labour force so as to provide an ostensibly authoritative historical account? In his opening paragraph of *Hard Times*, historian Studs Terkel writes:

> This is a memory book, rather than one of hard fact and precise statistic. In recalling an epoch, some thirty, forty years ago, my colleagues experienced pain, in some instances: exhiliaration, in others. Often it was a fusing of both. A hesitancy, at first, was followed by a flow of memories: long-ago hurts and small triumphs. Honors and humiliations. There was laughter too.
>
> Are they telling the truth? ... In their rememberings are their truths. The precise fact or the precise date is of small consequence. This is not a lawyer's brief nor an annotated sociological treatise. It is simply an attempt to get the story of the holocaust known as the Great Depression from an improvised battalion of survivors.[2]

Using oral histories, as Terkel did, is important in and of itself, but it is especially significant when there is not a comprehensive documentary record for different actors. While much of this book is highly annotated, engages sociological and historical scholarship, and draws from archival sources to the extent and availability possible, the "meat" of the story – the labour and immigration experiences of *pineros* and their families – is based on pain, recollection, and perseverance. Their personal narratives humanize and enliven the labour that takes place in the woods and, I hope, will lift the veil for those who care about the management of public lands. Ultimately, this account is an attempt to draw more systematic attention to immigrant forest workers and the multiple processes that produce their predicament, all with the aim of making the concerns of pineros more central to debates over natural resource management, labour standards, and immigration policy in the United States.

Acknowledgments

This book has benefited greatly from the advice, contributions, and support of countless individuals. First and foremost, I owe a debt of gratitude to the late Beverly Brown. My work would not have been possible without her steadfast encouragement and unwavering vision for social and environmental justice. I am also grateful to my colleagues at the University of California, Berkeley – in particular, Louise Fortmann, Sally Fairfax, and Kim Voss – who believed in the importance of this story from the beginning and who continue to be invaluable mentors. I am thankful to all who helped me during my stay in the Rogue Valley. I am especially grateful for the companionship and support provided by Shareen and Chris Vogel, Victoria and her family, my housemates Marjie and Jenny, Ann and Chris Muth, Katy and Duane Mallams, Terry Tuttle, Mark Chinn, Jose Montenegro, Milo Salgado, and the dedicated students of Club Latino at Rogue Community College.

While conducting my research, I benefited from the scholarship of Cassandra Moseley, who has written extensively on forestry in Oregon and has been generous with her feedback and data; from Vanessa Casanova and Josh McDaniel's research on forest workers in the US Southeast, which served as an important basis of comparison; from Jeff Romm's conceptualization of racial and resource reserves; from Jonathan Kusel and Katie Bagby's institutional work and scholarship on community forestry; and from Sarah Loose's popular education work on the issue of contingent labour. I also owe intellectual debts to Hal Hartzell Jr., who was prescient in conducting oral histories with forest workers in the 1970s and whose book on the Hoedads remains an essential reference; to Bill Robbins and Richard Rajala, who have examined the lives and livelihoods of loggers and whose work complements my contemporary focus on reforestation workers; and to Erasmo Gamboa's scholarship on *braceros* in the Pacific Northwest. Tom Knudson and Hector Amezcua's series on *pineros* in the *Sacramento Bee* was published as I was completing my research, and conversations with Knudson in particular helped me formulate ideas about drawing more public attention to the plight of forest workers. Finally, I am thankful to Dean Pihlstrom, who shared with me his collection of Associated Reforestation Contractors publications and who generously provided feedback on a portion of my manuscript.

Translating my research into book form has been a rewarding and challenging process. My colleagues in environmental analysis at Pitzer College – Paul Faulstich, Melinda Herrold-Menzies, and Susan Phillips – created a stimulating and supportive environment in which to pursue both research and teaching. Char Miller at Pomona College is a rich resource on the history of forestry in the United States, and I am thankful for the advice he provided regarding manuscript revision. I also benefited from the collegial support of faculty in many other disciplines, including my junior faculty compatriots and participants in the women-of-colour lunch group. More generally, I am grateful for Pitzer College's support for faculty research, which is evident in terms of not only funding opportunities, but also the general tone set from the highest levels down.

Outside of the Claremont Colleges, I have had wonderful mentors who shared timely advice over the years: Ruthie Gilmore, David Pellow, Donald Floyd, Geoff Mann, Devra Weber, Richard White, and Carl Wilmsen. I am also grateful for the professional advice and friendship of Sarah Loose, Vanessa Casanova, Jennifer Hughes, Bronwyn Leebaw, Andrew Jacobs, Catherine Allgor, David Biggs, Hong-Anh Ly, and Mary Braun. Thank you, too, to my good friends and colleagues from both inside and outside academia, who are too many to mention by name! They have provided moral support through countless emails, phone calls, and meals.

Thanks also to the anonymous reviewers who provided substantive feedback that helped improve this scholarship considerably. Randy Schmidt and Anna Eberhard Friedlander at UBC Press have been superb editors, and I am grateful for their support and skillful shepherding of the manuscript through the review and production process. Sandy Hamilton and Alan Jones at Pitzer College also provided support during this stage, and David Martinez lent a much-needed fresh pair of eyes to the manuscript before final submission. Programs and organizations that supported this research and its publication include the Social Science and Humanities Research Council of Canada, the Canadian Federation for the Humanities and Social Sciences, the Berkeley Human Rights Center, the Soroptimist Founder Region Fellowship, the Ford Community Forestry Research Fellowship, the Rural Sociological Society, the Morris K. Udall Foundation, and the University of California Institute for Mexico and the United States.

Finally, I am grateful for my loving and always supportive family: Appa and Amma, Ranima and Mama, Durga, my thambis and their partners, and my fabulous extended family. And I would never have finished this work without the patience, prodding, and companionship of Karthick Ramakrishnan. I am blessed to have him and our sons, Omji and Millan, in my life.

Pineros

1
Invisible Workers

The Pacific Northwest. In the national imaginary, this lush corner of the country conjures such iconic images of the natural world as towering firs, verdant rainforests, surging salmon, and spotted owls. The social actors who populate this imaginary are also linked to the natural realm, bearing testimony to their seeming inseparability. The logger is among the most prominent of these actors, with looming statues of Paul Bunyan still found in rural mill towns in Washington, Oregon, and northern California (see Figure 1.1). The independent spirit of the Western logger was perhaps best captured in Ken Kesey's popular epic, *Sometimes a Great Notion*, which became both a national bestseller and an Academy Award-nominated film. The novel and film chronicle the story of the stubborn Stampers, a family of generational loggers in the Pacific Northwest who draw the ire of townspeople because they are anti-union and refuse to co-operate in a local strike.[1] Folk songs such as those by Robert E. Swanson further exemplify America's romantic vision of the "lumberjack":

There's a life that is close unto nature,
where the soul is happy and free,
And you live by the brawn of your muscle –
ah, there is the life to suit me –
This job in a shipyard is lousy –
a paradise fit for a tramp.
So to Hell with a life in the city;
I'm off – to a logging camp![2]

Not all depictions of the logger's life, however, have been so heady. Scholars in particular have focused on the more sobering social aspects of timber extraction. In his account of a rural logging community in

Figure 1.1   Paul Bunyan statue in Shelton, WA. Photograph courtesy of Sarah K. Loose, 2004.

Coos Bay, Oregon, historian William Robbins reveals that loggers are often at the mercy of distant capital markets and left with little in exchange for their labour and regional resources.[3] Studies that focus on the economic displacement of timber workers in the western United States – due to technological improvements, mill closures, and the industry's relocation to the southern United States in the late 1970s – touch upon loggers' attachment to place and their reluctance to leave small towns in search of jobs elsewhere. These narratives vividly illustrate that poverty, alcoholism, and domestic abuse are all-too-common realities for many unemployed timber workers.[4]

The other renowned figure dominating our understanding of forests in the Pacific Northwest is that of the ardent environmentalist. High-profile activists like Julia Butterfly Hill, who lived for two years in a California redwood she named Luna, thrust local resistance to logging into the national and international limelight and became the subject of numerous pop-cultural references, ranging from an episode of *The Simpsons* to lyrics by the Red Hot Chili Peppers and Neil Young. Environmentalists have also been key players in forest politics through their effective use of litigation. Lawsuits spearheaded by activists over Forest Service violations of the Endangered Species Act and the National Forest Management Act famously led Judge William Dwyer to halt logging on spotted owl habitat in the Pacific Northwest. In 1994, President Bill Clinton implemented the Northwest Forest Plan, which reduced cutting on federal lands by 90 percent and fundamentally changed the region's political economy and management of national forests for decades to come. Despite their prevalence, however, popular images of loggers and environmentalists do not fully reflect the social reality of contemporary forest work (see Figures 1.2 and 1.3).

In the shadowy realms of the forest today, one will rarely find the white logger or the environmental activist. Rather, one may run into the likes of Juan Cabrera, an undocumented immigrant from Zacatecas, Mexico, who crossed the border at sixteen and has been tree planting on federal lands ever since.[5] One may chance upon Pedro Zamora Gómez, who was struck by an errant tree limb while thinning overgrown stands of fir, suffered a debilitating back injury, and was left to cope without health care. One may meet Santos Portillo, who started out planting saplings on rugged Forest Service terrain, later received his *papeles* (legal papers) through the 1986 amnesty, and is now a successful labour contractor running reforestation and fire crews of his own. Why are such stories of immigrant labour absent from mainstream conceptions of forest actors? Paul Bunyan and Julia Butterfly Hill belong in the woods; why is it that non-white immigrants are seen as an anomaly, a temporary blip on the forest management radar?

For well over a decade, Latino immigrants have constituted the majority of forest workers on both private and federal lands in the United States. Also known as *pineros*, they perform manually intensive activities such as piling and thinning brush, fuels reduction, pest control, and

Figure 1.2   Pineros thinning brush on Bureau of Land Management land, southern Oregon. Photograph by Brinda Sarathy, 2004.

reforestation, implementing many of the forest management techniques that produce "healthy forests." Immigrants from Mexico (both Latinos and Mixtec/indigenous groups), Central America, and parts of Southeast Asia also find work on public lands collecting *salal* (wild floral greens) and harvesting wild mushrooms.[6] These immigrants face unsafe working conditions and have little recourse in the face of labour violations and workplace exploitation. In the Rogue Valley of southern Oregon, which in many ways typifies the dramatic changes underway in forest work, the steady Latinization of the forest workforce has been underway since 1985. The 1980s witnessed a sharp rise in the entry of Mexican immigrants into tree planting, especially on federal lands. By the early 1990s, Latinization had spread to the ranks of forest labour contractors. Latino immigrants and their families are also fundamentally changing the broader demographics of rural residential communities in southern Oregon and in the Pacific Northwest in general.

Figure 1.3 Illustration by Sarah K. Loose, in consultation with Beverly A. Brown, published in *The Jefferson Center News* (Spring 2005): 5.

Still, the processes by which Latinos have come to dominate large portions of forest work remain unrecognized in debates over community forestry and forest management. Emerging in the early 1990s, the community forestry movement in the western United States was perceived by many as a potential solution to gridlock among environmentalists, industry players, and federal officials over the management of national forests.[7] The movement comprised various efforts by rural communities to gain control over the management of federal lands for both environmental and economic gains. Much of the community forestry scholarship in the United States examines collaborations, born out of environmental crisis or conflict, between community partners and the US Forest Service.[8] These various collaborations have been hailed for bringing disparate groups in rural communities together to discuss resource management issues, build community capacity, and provide opportunities for members to work as forest stewards.[9] Scholars have also portrayed such efforts as a foil to the top-down, technocratic approach of resource management developed in the Progressive Era of scientific forestry, and they celebrate collaboration in terms of its potential for reviving democracy and civic/ folk science.[10]

The primary stakeholders in resource management collaborations have been local representatives of timber companies, environmental activists, displaced loggers, and government agency officials. Given the presence of such diverse interest groups, the assumption is that everyone has a seat at the table. These "community-based" collaborations, however, exclude numerous individuals, including people who may not necessarily be full-time workers, local residents, or even citizens. The case of pineros raises the question of what it means to be a meaningful stakeholder when one's claims to basic rights such as pay for work, health care, and access to employment are compromised by immigration status. While decisions on how to manage forests affect all people who depend on natural resource-related work for their livelihood, involvement or participation in community forestry collaborations is not necessarily premised on one's labour. Many resource management collaborations do not acknowledge the plight of immigrant forest workers, let alone address their concerns. Thus, terminology like "stakeholders" and "participation" sometimes rings hollow. Stakeholders are not simply different groups with distinct perspectives who all have an equal voice at the table. Often, recognized stakeholders are groups and organizations who have access to resources and established relationships with decision makers. There are radical power imbalances in US society, which are also reflected in community forestry arenas. There are "people at stake" and there are "stakeholders" – the two are not always commensurable.

The scholarly privileging of white loggers, logging company officials, environmental activists, and federal land management agencies reinforces the elision of immigrants from forest management. Prominent books such as *Anatomy of a Conflict, Deadfall, Community and the Northwest Logger,* and *Hard Times in Paradise* represent white male US citizens as the typical face of forestry in the Pacific Northwest.[11] Similarly, federal policies – such as President Clinton's Northwest Forest Plan and President G.W. Bush's Healthy Forests Restoration Act – have been silent on the issue of forest labour. Together, these oversights constitute a series of omissions about how and by whom federal lands are managed.

Representations of white loggers and environmentalists as the *only* agents in the realm of forests are deeply problematic. Yet the racialization of forests and other "natural" landscapes as primarily "white spaces" is not new. The omission of people of colour from such arenas extends back to beliefs about the manifest destiny of white pioneers, embodied in Frederick Jackson Turner's 1893 frontier thesis and in the forceful removal of Native peoples from their territories in order to create untouched wilderness for the enjoyment of white tourists.[12] Turner argued that the "frontier" – as the border shaped by the constant move westward and the "meeting point between savagery and civilization" – was the crucible in which Americans and Americanism were forged. Characteristics of the quintessential American, "that coarseness and strength combined with acuteness and acquisitiveness; that practical inventive turn of mind, quick to find expedients; that masterful grasp of material things ... that restless, nervous energy; that dominant individualism," were all attributed to the influence of the frontier.[13] With white males depicted as the only relevant actors on Turner's frontier, the agency of other groups was written out. Such a racist, sexist, and imperialist portrayal of free and western lands ignored the presence of Native Americans and their eventual and bloody subjugation, as well as already settled Hispano lands. Democracy and freedom, as Turner asserts, were not simply intrinsic to the frontier but were premised on the active oppression of other groups.[14]

The notion of forests as "white spaces" may thus be understood as a historical product of racialized violence and white supremacy in the American West. In the nineteenth century, for example, the "racial line was drawn along European and non-European lines," and non-white groups (Native Americans, Mexicans, Blacks, Chinese, and Japanese) were ordered both in relation to one another and, inferiorly, in relation to whites. Prominent legislation that institutionalized white supremacy and gave white male citizens privileged access to land include the Chinese Exclusion Act (1889), the Alien Land Act (1913), and the Johnson-Reed Act to exclude Japanese labourers (1924).[15]

Today, the consequences of such historic exclusion are partially seen in the lack of minorities using forests for recreation and their limited representation within federal agencies such as the US National Park Service and the US Forest Service.[16] Scholar Jeff Romm argues that the whiteness of forest spaces may be understood as a product of simultaneous processes of social and resource restraints. Romm compares historic developments in the American South and American West. Freed slaves had to buy their own land after the Civil War and Reconstruction, while whites were encouraged to move out West and were given free land to homestead. The immobilization of the black South is thus inextricably linked to the opening of a white West. Romm also extends the analysis of social and resource restraints to the creation of national forest reserves during the Progressive Era. The development of national forests increased the value of land outside reserves. As more people became concentrated on that land, wages in these areas decreased. Prior to the creation of forest reserves, the public domain was disposed of, or given away in the case of the Homestead Act of 1862, to those who "improved" land through clearing it for settlement. Racial segregation, moreover, ensured that ownership of remaining private lands went to whites.

The western United States was partly formed through a system of race and resource-based reserves that maintained an exclusive order of white governance whereby people who lived in and depended on the forests, if they were not white, were overlooked by forest policies altogether. Over the course of the nineteenth century, Native Americans were driven off their homelands and fenced into reserves. Race-based immigration policies were used to secure pools of unprotected, low-wage immigrant labour for local farms and fisheries. Later, New Deal protections did not extend to agricultural workers. World War II initiatives like the Bracero Program institutionalized non-white farm labour and created a de facto system of social and spatial restraints through which people were confined to particular (often agricultural) places.[17]

Of course, one can justifiably argue that federal agencies such as the National Park Service, Forest Service, and Bureau of Land Management were never established to *intentionally* exclude people of colour. Moreover, the language of federal forest policies is race-neutral and addresses neither white nor non-white workers explicitly. The assertion of forests as white spaces is thus not to suggest that the federal government sought outright to exclude minorities from these lands. Federal policies, however, were implicitly racialized, given historic, political, and economic power dynamics (who wielded power, and with what priorities), and have contributed to institutionalized racism in both forest work and recreation today.[18]

Besides scholarship on forestry, pineros are invisible in research on immigrant settlement and labour incorporation. A number of studies have already shown that immigrants, especially from Mexico, are now settling outside traditional destinations such as California, Texas, and Arizona. A variety of factors have contributed to this shift, including anti-immigrant policies in border states and increases in labour competition among low-wage workers.[19] Since the 1990s, Mexican immigration has shifted towards new states, and there are now significant Latino settlements in the southern United States, the Midwest, and parts of the Pacific Northwest.[20] Still, research on immigrant labour remains predominantly focused on national trends and on socio-economic outcomes in urban areas.[21] When scholars do pay attention to immigrant labour in rural areas, the focus is primarily on farm workers, not forest workers.

More recently, immigrant forest workers have started to receive some attention from scholars and journalists. In November 2005, the *Sacramento Bee* documented the plight of pineros across the United States in an award-winning investigative series that eventually led to congressional hearings on guest workers on federal lands.[22] Ironically, these hearings did little to address the labour abuses faced by pineros in the Pacific Northwest, most of whom are not guest workers or otherwise documented.[23] Immigrant forest workers have also received periodic attention in state and national newspapers, with feature stories on fire fighters and floral-greens harvesters.[24] These articles raise troubling questions about the management of public lands by a vulnerable workforce of low-wage immigrants.

While important, most of the extant scholarship on Latino forest workers has focused on plantation forestry in the US South[25] or on non-timber forest product harvesters in the Pacific West.[26] Industrial forestry on private lands in the South is often subsidized by state incentives to pulp-and-paper corporations and is a for-profit operation. It differs starkly from the ecosystem management work on federal lands in the West, which is funded primarily through congressional appropriations to the Forest Service budget. In Alabama, Anglo labour contractors recruit workers from Mexico and Latin America through kinship networks and via a system of H-2B guest worker visas.[27] Similar patterns of worker recruitment, through H-2A visas, are found for Christmas tree producers in North Carolina.[28]

Taken together, these cases do not address the extensive use of immigrant workers on public lands. In contrast to private plantation forestry in the South, which depends primarily on labour recruitment through a system of guest worker visas, or tree planting in Canada, which relies mostly on native-born college students,[29] undocumented immigrants constitute the majority of reforestation workers on federal lands in the western United States.[30] Whereas environmental activism on public lands has centred on logging, fire, and forest health, the issue of immigrant labour exploitation continues to be overlooked. This is particularly interesting because it points to a phenomenon whereby the labour force on private lands is actually *more* regulated (at least in the Southeast) – through a system of H-2B and H-2A guest worker visas – than the workforce on federal lands in the West, which remains largely unregulated and undocumented. While a significant number of workers on private lands are also undocumented, a predominant and sustained focus of immigration policies in only the private sector helps perpetuate ignorance about immigrants working in the public realm.

Overall, then, current scholarship has yet to systematically address deeper questions regarding the arrival, settlement, and continued exploitation of immigrant forest workers: How, in fact, did Latino immigrants come to manage national forests? How do federal agencies perpetuate systems of labour exploitation? What are the working conditions of pineros today? Are the concerns and perspectives of immigrant workers represented in debates and policy considerations about forest management? These are some of the central questions guiding the analyses in this book.

What follows is a social history of Latino immigrants managing forests on federal lands. The area of a forest that grows in the shade of the canopy layer is called the "understory." Most manually intensive forest work takes place in this shadowy realm. Latino immigrants are thus both figuratively and literally hidden in the understory. This book not only widens the scope of environmental history in Oregon; it also sheds light on the broader dynamics of immigrant settlement and incorporation in parts of the rural West. I address four important sets of questions:

1   How has the forest workforce changed on federal lands in the western United States? What work is done in the forest and by what groups of workers?
2   What factors have accounted for the rapid Latinization of the forestry workforce in southern Oregon? What accounts for the exploitation of thousands of undocumented immigrants on federal lands?
3   How does the restructuring of work on federal lands contribute to rural Latino immigration in the Pacific Northwest?
4   What are the normative, political, and policy implications of using a highly marginalized workforce to implement forest management policies?

The Rogue Valley is a key site for studying the Latinization of forest labour for several reasons. In recent years, more procurement dollars have been spent on forest management work in southern Oregon than in other parts of the state, and contractors from this region have won a greater share of federal contracts in forestry.[31] This area has seen a rapid rise in the number of settled Latino contractors and forest workers since the early 1990s. These contractors and their crews work on federal lands throughout southern Oregon and northern California, including the Umpqua National Forest, the Rogue River-Siskiyou National Forest, and various Bureau of Land Management lands. Finally, 53 percent of the land in the Rogue Valley is federally owned, a proportion that is similar to the rest of Oregon.[32] This high ratio makes federal lands the largest sites of labour exploitation, and the federal government among the largest employers (albeit indirectly), of undocumented workers in the United States.

Analytic Frames
The changing face of forestry in the Pacific Northwest can be viewed through at least three different theoretical lenses. First and foremost, this is a story about the changing racial composition of manual work in the woods, from a predominantly native-born and white labour force to a largely undocumented Latino population. Second, it is about differential power dynamics within ethnic groups and the ways in which kinship ties affect contractor-employee relations in the growing arena of contingent labour. Finally, one can understand the shifting demographics of, and related consequences for, distinct groups of forest workers as a product of the overlooked intersections between immigration, land management, and labour policies.

Geoff Mann's scholarship helps theorize the racialization of pineros in the Pacific Northwest. Mann examines the rhetorical use of the term "wage slaves," which is employed by both white workers and Latino forest workers in the Northwest. Historically, the "wage slavery" metaphor was used primarily to protest the conditions faced by white workers in the West, who felt economically threatened by immigrant labour. Mann argues, however, that today, Latino immigrants use this metaphor to highlight their own exploitation and to demand equal status with white workers. Whites, on the other hand, use "wage slavery" to refer to the poor treatment of immigrant workers and also to the unfair low-wage bidding system for Forest Service contracts. In effect, the substandard wages given to immigrants are perceived as eroding the position of white workers. White workers thus articulate a class-specific form of anti-immigrant sentiment in seeing the interests of "all workers" – read as white – with a legitimate claim to employment in the sector as damaged by undocumented Latino labour ("illegal aliens"). Moreover, this leads to all Latino workers being perceived as undocumented.[33]

Mann's distinction between white and Latino workers is ultimately complicated by the presence of contractors who are also primarily Latino. Thus, while a representation of Latinos as a homogeneous and "subjugated" racial category in relation to whites has some truth, it does not fully address the issue of varied power dynamics and intra-racial exploitation among Latinos. As this study will highlight, immigrant networks and family ties often exacerbate the labour subjugation and wage slavery of undocumented relatives by Latino contractors and foremen. In effect, it is not clear whether Latinos actually compete with whites for the same kind of work. The question of wage slavery may not necessarily pertain to an Anglo-Latino labour relationship, but, rather, to dynamics of intra-ethnic labour exploitation between Latinos themselves.

This raises the question of whether social networks and kinship ties serve as positive forms of social capital. Put another way, are networks of friends and family helpful to new immigrants? In their study of Cuban and Mexican immigrants in the United States, sociologists Alejandro Portes and Robert Bach claim that ethnic enclaves provide spaces in which immigrants can move up the social ladder, despite discrimination from outside society. Immigrants benefit by getting jobs through friends and relatives, have the opportunity to learn the ropes, and progress in terms of social mobility.[34] However, others have rightly argued that ethnic enclaves, and particularly relations of social obligation, are not necessarily beneficial to all immigrants. Enclaves may favour immigrant entrepreneurs but not necessarily low-wage immigrant workers. Scholar Cecilia Menjívar illustrates that social networks cannot always be taken as positives. Rather, the nature of government policies towards immigrants, the local political-economic context for new arrivals, and the presence of organizations serving immigrants all influence the ways in which networks may either hinder or help immigrants in new areas.[35]

Additionally, most scholars have examined the marginalization of immigrant workers using a dichotomous, primary/secondary approach to labour markets.[36] Differences between a primary labour market, with secure, high-paying jobs, and a secondary labour market of low-wage and less secure jobs are used to explain the less-than-desirable occupational position of non-white immigrants.[37] While differentiating between labour markets is helpful, such dichotomization can overlook the politics and policies creating such markets in the first place. To this end, Brendan Sweeney emphasizes economic, political, and sociocultural factors shaping labour market segmentation in Ontario's tree-planting industry. He convincingly shows that a host of causes, including government policies, changes to the forest products industry, and cultural stigma associated with reforestation, all contributed to the marginalization of tree planters in Canada.[38]

I employ a similar causal analysis to understand the marginal position of pineros. While Latino forest workers are certainly relegated to the confines of a secondary labour market, their marginality has been shaped by multiple factors including changes to federal land management and immigration policies, and new labour opportunities available to native-born whites. I thus use the concept of a "segmented labour force" (versus a "segmented labour market") to refer to internal segmentation within the forest management industry, and specifically the distinctions in legal status between documented Latino contractors and undocumented workers. As I will show, not all Latinos in forestry are equally marginalized.

Overview of Chapters
Chapter 2 delves into the recent history of reforestation efforts on public lands in Oregon. I highlight reforestation's intimate connection with large-scale logging operations and how both processes constitute activities, whether destructive or restorative, in which people strive to utilize and manage natural landscapes and resources. I then document the presence of a diverse reforestation workforce in Oregon that emerged in the 1970s, including numerous contractor-run crews and hippie-based co-operatives. I examine the experiences of these different groups and also focus on the more general competition between tree-planting co-operatives and contractors. This chapter closes with the argument that the decline of reforestation co-operatives that began in the 1980s was largely tied to changes in labour law (which made co-operatives unable to compete with contractors) and workers' pursuit of employment opportunities in other sectors. As a result, reforestation work became available to larger numbers of immigrant workers in the late 1970s and early 1980s.

In Chapter 3, I present a comprehensive analysis of the Latinization of forest work in the Rogue Valley. Here, I rely on oral histories and archival material to construct a narrative of Latinos' involvement in forest work and immigrants' broader experiences of settling in the area during the early 1980s. The chapter explores group experiences of isolation and racism, as well as the phenomenon of Latinization, which I detail in three parts. First, I look at the movement of immigrant workers "from the pears to the pines" and, correspondingly, "from the field to the forest." I then document the transition of many Latinos from forest workers to forest labour contractors in the late 1980s and early 1990s. Finally, I explore the broader Latinization of the settled population of the Rogue Valley and show that immigrants are significantly changing the demographic composition of the region.

In Chapter 4, I analyze the varying labour marginality of Anglo loggers, tree planters, and pineros. I focus on aspects of workers' economic and social marginality and examine their labour conditions, coverage by the media, and political visibility. Working conditions include safety on the job, wages, access to benefits, protection from labour violations, and the structure of labour arrangements. I also examine media coverage of forest workers in national and local newspapers, and identify policies and organizations that benefit or represent the concerns of particular groups of forest workers. Ultimately, this chapter highlights the idea that pineros are the most marginal of the three groups of forest workers and that vulnerable legal status and limited English proficiency contribute to their disadvantaged positions.

Chapter 5 moves beyond the labour marginality of pineros to explore how immigrants organize to confront their exclusion and exploitation. Although pineros' lack of legal status and limited English proficiency may hinder their ability to mobilize, groups elsewhere have been able to overcome similar barriers. I thus begin by examining the role of advocacy organizations such as the United Farm Workers in California and the Willamette Valley Immigration Project in Oregon during the early 1970s. I compare the organizing success of these groups with their less activist counterparts in the Rogue Valley and find that the era in which organizations emerged matters. Both the United Farm Workers and the Willamette Valley Immigration Project (which later became Pineros y Campesinos Unidos del Noroeste, Oregon's only farm worker union) were founded, and successfully mobilized immigrant workers, prior to the 1980s. By contrast, the organizations that serve immigrants in the Rogue Valley, which have been far less radical overall, were all established after 1985. This pattern suggests that there were windows of opportunity prior to the 1980s that made organizing among immigrant farm and forest workers in California and Oregon more likely and more successful than in the years that followed.

In Chapter 6, I conclude the book with a discussion of the ethical and policy implications of having a large number of marginal immigrants managing federal lands. I also suggest ways to improve the working conditions of pineros and increase their involvement in debates over resource management, labour standards, and immigration policy. Finally, I call for a re-envisioning of the history of forest work such that the social realities of Oregon's understory may be more accurately represented.

2
Cutting and Planting

Urbanites today generally associate the national forests of the Pacific Northwest with wildlife, outdoor recreation, and leisure. Yet for almost half a century, federal lands in the Northwest supplied the majority of timber for a nation in the throes of a postwar construction boom. Locals commonly referred to national forests as the "nation's woodlot," and timber was hailed as "king" in Oregon and Washington. The trees that fill the lush Northwest are by no means pristine and untouched original forests. Rather, they are second- or third-growth species: as saplings planted by reforestation workers over four decades ago, these trees grew in the aftermath of one or more timber harvests. These trees also represent the fruit of the federal government's emphasis on reforestation in a number of statutes from the mid-1900s on. The Forest Service channeled resources to state agencies to promote reforestation practices on both private and public lands, while the timber industry began experimenting with tree plantations in the 1930s. Later, as national forests became key suppliers of timber in the postwar period and new state Forest Practices Acts were passed, the reforestation industry took off. During the early 1970s, tree planters were primarily native-born and white, but this gradually changed as Latino immigrants entered the labour force in the mid-1980s.

We have rich historical and sociological accounts about loggers, but we know little about the people who planted the trees so emblematic of the Northwest. This pattern of oversight continues, with limited attention paid to pineros today. How were groups recruited to work in the woods, and what were working conditions like? How was the workforce organized, and how has it changed over the years? To better appreciate the trees in our forests today, and the people who planted them, we might take a step back to examine these questions more deeply.

Lumbering in the Pacific Northwest

As the population of the United States expanded rapidly during the 1800s, so too did the nation's demand for both wood and cleared land for agriculture. The number of residents increased from 5.3 million at the start of the nineteenth century to over 76 million in 1900.[1] During the same period, deforestation skyrocketed from less than half a million board feet (the cubic unit of measure for lumber, equal to one square foot by one inch thick) of lumber to approximately 45 million board feet.[2] The prodigious building of railroads, coupled with the introduction of the steam donkey and overhead yarding techniques, launched lumber as "big business."[3] While Maine, Pennsylvania, and New York initially led the nation in lumber production, it was the cut-and-run logging of vast white pine forests in the Great Lakes region through the mid-1800s that set the norm for the "nomadic timber baron."[4] In 1951, William B. Greeley (chief of the Forest Service from 1920 to 1928) described the unbridled resource exploitation of the timber barons with the following observation:

> The drive of large capital investments for speed and profit brought about the rapid skinning of enormous areas of timberland. Twenty years, even less, became the common lifetime of a sawmill. Then – dismantle, junk, and move on. Not only did lumbering perforce become a nomadic industry; it became an industry with no permanent interest in the land. A logged-off section was in the same category as a junked sawmill – to be sold for what it might bring, or abandoned and forgotten ... This devil-take-it attitude toward the land was strengthened by the common belief that most of it would soon be in cultivation anyway.[5]

Many factors contributed to the plundering of timber in the nineteenth century. The General Land Office, within the Department of Interior, was infamous for its loose and often corrupt administration of land laws and helped to promote the disposition and privatization of the public domain. In some cases, General Land Office-appointed forest rangers in the late 1890s were complicit in land frauds committed by their associates or accepted money to help homesteaders obtain preliminary title to forest land that was then sold to speculators or timber companies. In 1903, the *New York Times* reported the government's discovery of a "stupendous land graft ring" based out of San Francisco. The ring, which had been operating in every well-timbered area on the Pacific Slope, purportedly duped state land officials, making them partners in their scheme, and "maintained in the General Land Office at Washington, agents, whose duties were to leak information about proposed reserves and other profitable matters, and, by use of money, influenced the placing of reserved boundaries to its own interest."[6]

The political economy of the lumber industry itself fueled rapid and reckless extraction. Early loggers and sawmill operators felled forests with little thought to future timber production and ravenously followed the timber supply, from the East Coast and the Lake States on down to the South. Until 1860, the eastern seaboard led timber production, with New York accounting for 30 percent of the nation's lumber in 1839. Combined with production from Maine, Pennsylvania, and New England states, the East accounted for 63 percent of total wood production.[7]

Increasing lumber demand eventually led the timber industry to shift to the Lake States of Michigan and Wisconsin. Overcapitalized firms, having invested heavily in timberlands and mill infrastructure, were pressed into a relentless cycle of production to meet debt and interest payments. Such situations often led to the panicked liquidation of stands, further glutting markets and pushing down prices.[8] Between 1869 and 1889, timber production in Michigan was five times that of New York. As the seemingly endless supply of Great Lakes white pine finally started to dwindle in the early 1880s, lumber barons from Minnesota and Wisconsin moved operations to the South and rapidly proceeded to deplete forest resources in states like Georgia and Alabama.

Coming thirty years after the shift of timber production from the East to the Lake States, more advanced and efficient technology facilitated the decimation of southern forests. The use of steam engines and circular saws, for example, replaced many of the older waterwheel sawmills, which often worked in unison with gristmills.[9] As a result, production of saw timber in the South went from 1.6 billion board feet in 1880 to an estimated 15.4 billion board feet in 1920.[10] Southern states also lobbied to make their natural resources available to industry, and the region's landholding patterns allowed for the purchase and consolidation of millions of acres into private ownership. At one point, a mere 925 individuals acquired nearly 46.6 million acres of standing timber in the South, over half of the region's supply.[11] By 1920, the South had reached its peak in timber production and would soon be overtaken by the Pacific Northwest.

By the end of the nineteenth century, all eyes were looking westward to the last remaining stands of untouched timber. Contemporaneous with the rise of the Forest Service in 1905, the American West saw a rapid expansion in lumber production on private lands. In 1900, a group of

Lake States lumbermen led by Frederick Weyerhaeuser purchased 900,000 acres of forested land from the Northern Pacific Railway Company. Weyerhaeuser went on to acquire and hold more stumpage (standing timber viewed as a commodity) over the ensuing years, while many smaller timber operators had to cut desperately just to stay in business. Between 1904 and 1939, spells of intermittent overproduction and low timber prices led to the combined harvest of an estimated 320.5 billion board feet from Washington, Oregon, and parts of California and Nevada.[12]

The species of choice in western Oregon was the mighty Douglas fir (*Pseudotsuga menziesii*), found in the mild and damp forests west of the Cascades.[13] Lumber manufacturers favoured Douglas fir because it regularly grew to heights of 175 to 200 feet, almost always with a straight, non-branching trunk, and had dense and straight-grained wood ideal for house-framing, floorboards, and veneers. Given the tree's size, shape, and girth, Douglas fir also yielded more board feet of timber per acre than any other species.[14] In central Oregon's more arid Deschutes-Klamath district, stands of ponderosa pine (*Pinus ponderosa*) made for prime timber until the early 1940s. However, years of rapid timber harvests and concerted efforts at fire suppression soon changed the open-stand ecosystem of the pines, leaving behind dense thickets of fire-prone fir.[15]

For the first three-quarters of the twentieth century, clear-cutting was the dominant mode of production in the forests of the Pacific Northwest. Newly adopted steam-powered ground-lead systems (the steam donkey) yarded large swaths of timber over rough terrain but also foreclosed the option of selectively saving younger trees.[16]

By 1915, timber extraction had transitioned to overhead techniques, including high lead and skidder methods that used aerial cables to transport logs. These systems necessitated complete clear-cutting to allow "rapid shifting of lines and unimpeded passage of logs to the landing." Small trees that had escaped logging were typically uprooted during the yarding operation as logs were quickly rigged and transported by massive skidders.[17] Technocratic foresters zealously welcomed such industrial developments and ushered in the practice of total clear-cutting.

If technology underpinned clear-cutting as the primary mode of extraction, scientific rationale reinforced its acceptance among foresters. Research by E.T. Allen in 1901 showed that the Douglas fir germinated best on exposed mineral soil open to full sunlight.[18] Foresters used the shade-intolerant characteristic of Douglas fir to justify clear-cutting because it resulted in open stands, allowed slash burning to eliminate

duff (the layer of organic matter covering the forest floor), and hypo-
thetically created the ideal reproductive environment for the species.
Foresters' perceptions of old-growth forests as decadent and unproductive
also lent support to clear-cutting as the preferred technique. Finally,
some foresters accepted even more questionable ideas about Douglas fir
regeneration that conformed with, and further promoted, clear-cutting.
The 1913 "seed storage" theory propounded by J.V. Hoffman, director
of the US Forest Service's Wind River field station, is a case in point.
Hoffman theorized that Douglas fir seedlings had the capacity to remain
viable for long periods of time while buried in the duff layer on the forest floor. Once exposed to sunlight (as a result of canopy clearing due
to either logging or forest fires), these seeds would germinate. Although
Hoffman's theory was discredited by the 1930s, his "interpretation of
Douglas fir reproduction legitimated unregulated clear-cutting and by
extension provided coastal operators with a scientific rebuttal to use
against advocates of cutting practice regulation."[19]

Although private timber interests were the prime culprits in the decimation of western forests in the early twentieth century, the Forest
Service also played a role. Throughout the Progressive Era, American
foresters were driven in their quest for rational planning and guided by
the paradigm of scientific forestry. Originally developed in Europe,
"sustained yield" or "scientific forestry" addressed the conditions of
"scarcity, stability and certainty" in nineteenth-century Germany.[20]
Given the scarcity of forestland, planned harvesting and planting cycles
offered a way to meet the demand for wood. German foresters thus
managed their forest stands intensively and for maximum timber production. Bernard Fernow, head of the US Division of Forestry, subsequently brought German sustained-yield principles to the United States.
In 1905, Gifford Pinchot, as first chief of the US Forest Service, formally
institutionalized scientific forestry, and it soon became the reigning
method of managing forests on federal land.

At its core, scientific forestry had two key principles: forests are a renewable crop that can and should supply the nation with a continuous
supply of timber, and forests are best managed by a technocracy of experts
trained in silviculture. Scientific foresters decided which trees would be
harvested and which would remain standing as seed trees. They decided
what species to replant, where, how, and how many. Yet the gospel of
scientific forestry was fraught with contradictions. Ecologically, the forest
management techniques applied by foresters were poorly suited to the
arid and expansive landscapes of the western United States. Politically,

foresters gave in to pressures exerted by timber interests and accelerated the harvest cycle by twenty to thirty years, defying even their own sustained-yield management prescriptions.

Environmental historian Nancy Langston reveals how, in the Blue Mountains of eastern Oregon during the 1920s, foresters ended up liquidating old-growth ponderosa pines much faster than the forest could replace them:

> Federal foresters came to the Blues with a vision of working with wild nature to make it perfect – efficient, orderly, and useful ... Yet in trying to make it green and productive, they ended up making it sterile ... Foresters destroyed the forests, not in spite of their best intentions, but *because* of them – precisely because foresters' ideas of what was good for the forest were based on an ideal of deliberately transforming nature to serve industrial capitalism.[21]

Foresters had an unshakeable faith in sustained yield, but by the 1920s, many had come to realize that industry's drive for profit had compromised the principle.

Clear-cutting ideology dominated as the mode of production in forestry, but it did not go wholly undisputed. In the midst of the 1930s economic crisis, for example, lumber operators used Caterpillar tractors in order to selectively log trees to reduce labour costs and address restricted markets. This practice left considerable amounts of slash on the ground (a fire hazard), along with poor-quality standing trees in partially cut areas. Forest Service officials became divided over the technique, with the Pacific Northwest Forest Experiment Station's Thornton Munger committed to clear-cutting, while regional forester C.J. Buck viewed selective logging as a remedy for waste and overproduction.[22] In the long run, of course, clear-cutting prevailed and would continue to outpace any serious effort at active reforestation on private and federal lands for the next four decades.

The Logging of Federal Lands

In 1942, wood became classified as a "critical war material." Second in demand only to steel, it was an indispensible resource for numerous types of infrastructure including hangars, scaffolding, airplanes, wharves, bridges, pontoons, railway ties, telephone poles, mine props, shipping containers, and air-raid shelters.[23] As a result, the War Production Board and the Forest Service worked in unison to increase timber production on private and public land. In his address to the Washington Section of the Society of American Foresters, Arthur Upson, chief of the Lumber and Lumber Products Branch of the War Productions Board, captured the essence of "all-out lumber production for all-out war." He made clear that any effort to regulate cutting was to be put on hold. The war trumped conservation of resources and justified the rampant and unregulated harvesting of timber from federal and private arenas:

> Even though it is against my teachings and my judgment if this were peace time, I think we will have to cut our forests heavily if lumber is to contribute what it should to winning the war; and the war cannot be won quickly without lumber. Therefore, I do not think that now is the time to consider any type of rigid public control of cutting. [24]

In the years following World War II, regulations on cutting remained limited and timber harvests from national forests rose exponentially (see Figure 2.1). The Forest Service's transition from custodial responsibilities to intensive timber production resulted largely from the war and postwar economy, and from policies such as those established by the Sustained Yield Forest Management Act of 1944. Aimed "to promote the stability of forest industries, of employment, of communities and of taxable forest wealth through continuous supplies of timber," the act embodied efforts to reshape forest policy in line with industry's needs to stabilize trade in forest products.[25] The legislation allowed the Forest Service to combine private and public land into cooperatively managed sustained-yield units and permitted similar arrangements on federal lands where communities were dependent on timber for their livelihood. While ostensibly aimed at stabilizing both communities and the lumber industry, the promise of sustained-yield legislation never materialized. As William Robbins asserts, "the health of the market remained the primary determinant of harvesting rates on private lands."[26] Overall, timber sales on national forests rose 238 percent between 1939 and 1945, representing a jump from 1.3 billion board feet to 3.1 billion board feet.[27]

The Pacific Northwest soon became the nation's woodlot, with a regional economy more dependent on timber production than anywhere else in the United States. In 1957, Oregon and Washington alone housed nearly 40 percent of the country's standing timber and accounted for more than 30 percent of its lumber supply.[28] Spurred by a nationwide building boom, the postwar years brought prosperity to the Northwest and squelched most concerns, raised during the 1930s, about rampant resource depletion and the need for planned stewardship of timberlands. In the ensuing decades, the Forest Service implemented resource policies

Figure 2.1

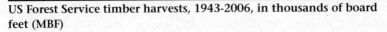

US Forest Service timber harvests, 1943-2006, in thousands of board feet (MBF)

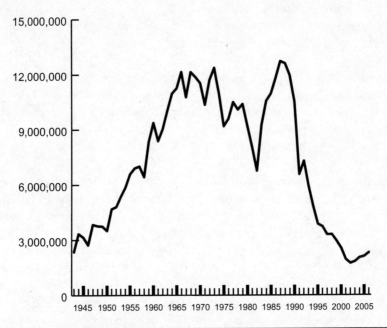

Source: US Forest Service, *US Forest Service Forest Management – Cut and Sold Reports* (Washington, DC: US Department of Agriculture, Forest Service, 2010), http://www.fs.fed.us/forestmanagement/reports/sold-harvest/cut-sold.shtml.

to manage land primarily for production and steadily increased the allowable cut from national forests. By the late 1960s, over twelve billion board feet of timber were being harvested from national forests *annually*. The heyday of cutting on federal lands continued into the early 1980s, and Region 6 of the Forest Service (composed of Washington and Oregon) accounted for the lion's share of timber harvests on national forests until the early 1990s (see Figure 2.2).[29]

Throughout this period, Oregon was easily the most timber-productive and, consequently, timber-dependent state in the country. Timber harvests accelerated from 1934 on, and by 1947, the state was home to more than two thousand logging operations "with a combined payroll exceeding all other employment."[30] While private timber accounted for a large portion of harvests prior to World War II, Oregon's federal lands supplied

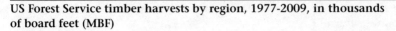

Figure 2.2

US Forest Service timber harvests by region, 1977-2009, in thousands of board feet (MBF)

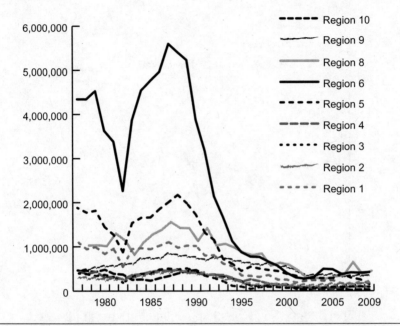

Source: US Forest Service, *US Forest Service Forest Management – Cut and Sold Reports* (Washington, DC: US Department of Agriculture, Forest Service, 2010), http://www.fs.fed.us/forestmanagement/reports/sold-harvest/cut-sold.shtml.

the bulk of production in the postwar years (see Figure 2.3). By the late 1960s, the timber industry employed one of every eleven Oregon workers, and the state led the nation in timber production until the mid-1980s.[31] Postwar timber harvests were 9.3 billion board feet in 1956, 9.4 billion in 1962, and 8.7 billion in 1986, with over 220 million board feet coming out of southern Oregon alone.[32]. By contrast, Washington, which had the nation's largest timber harvests prior to World War II, reached a postwar high of 3.9 billion board feet in 1965.[33] Overall, Oregon's national forests accounted for approximately 27 percent of all Forest Service timber harvests between 1978 and 2006 (see Figure 2.4). The felling of trees on federal lands in Oregon was not only integral to the state's economy; it also constituted more than a quarter of national timber production.

Figure 2.3

Changing shares of timber harvests in Oregon, 1962-2004, in thousands of board feet (MBF)

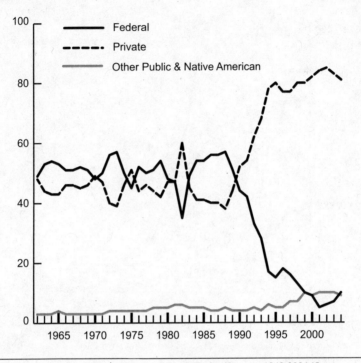

Source: Alicia Andrews and Kristin Kutara, *Oregon's Timber Harvests: 1849-2004* (Oregon Department of Forestry, 2005), 7, http://www.oregon.gov/ODF/STATE_FORESTS/FRP/docs/OregonsTimberHarvests.pdf?ga=t.

Reforestation by Private Industry

In response to threats of federal government intervention, west coast lumbermen began to take steps towards self-regulating forest practices as early as 1933. Private industry found an ally in Forest Service Chief William Greeley. In contrast to Gifford Pinchot's avid support of federal oversight, Greeley favoured a more voluntary, industry-regulated approach to forest management. Under Franklin D. Roosevelt's National Recovery Administration, the US Forest Service and private timber interests together actively drafted language on forest practices for the National Recovery Administration Lumber Code. Greeley recounts his own participation in this effort:

Figure 2.4

US Forest Service timber harvests, 1977-2009, in thousands of board feet (MBF)

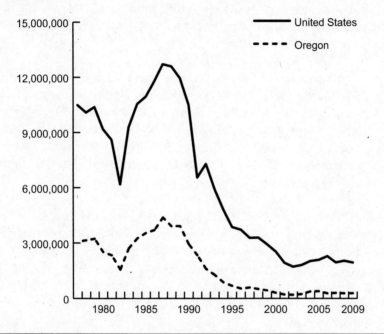

Source: US Forest Service, *US Forest Service Forest Management – Cut and Sold Reports* (Washington, DC: US Department of Agriculture, Forest Service, 2010), http://www.fs.fed.us/ forestmanagement/reports/sold-harvest/cut-sold.shtml.

"President Roosevelt," I said, "is almost certain to want something in this code on *forestry*. Let's beat him to the draw. It will help us get the rest." It was agreed. I was commissioned to draft a clause which would commit the industry to a reasonable program of forest conservation. So it came to pass, after all the discussions and compromises. Article X of the Lumber Code committed the industry to leaving "its cutover lands in good condition for reforestation." The Forest Service concurred. In a final session at the White House the President beamed his approval ... We left the Capitol with shiny new halos on our heads."[34]

From the start, the industry dictated National Recovery Administration attempts to regulate timber harvesting. In the Pacific Northwest, lumbermen and loggers formed a Joint Committee on Forest Conservation and formulated rules for administering Article X in the Douglas fir region. Voting members on the committee were drawn exclusively from industry, while government officials (state and regional foresters) were relegated to non-voting, advisory positions. With complete control over the rule-making process, timber operators crafted vague rules that were rife with qualifications and loopholes:

> Young growth and trees reserved as seed sources were thus to be protected "as far as practicable" or "so far as feasible" from logging damage. Clear-cuts were to be no further than a quarter-mile from reserved timber of sufficient quantity that it might "reasonably be expected to furnish an adequate source of seed." ... Nothing in Article X rules drafted by the Joint Committee obligated Douglas fir region operators to anything beyond the normal conduct of logging.[35]

Despite the corporations' nod towards reforestation, industry self-regulation ultimately meant that plans for additional cutting outweighed any serious considerations of replanting. Reforestation, after all, was an expensive prospect that flew in the face of economic reality during the Great Depression. In Washington in 1938, for example, "the cost of re-forestation was $12 to $15 per acre against the tax sale purchase of timber at $4 an acre."[36]

Gradually, however, incentives emerged to encourage reforestation on private and federal lands. The passage of the Knutson-Vandenberg Act in 1930 authorized federal funds for reforestation nurseries and provided additional means for tree planting in national forests. Revised land tax laws, imposing a levy on timber only once it was cut, also removed the incentive for timber companies to liquidate standing trees and sell off land without concern for replanting. Publications that promoted sustained-yield forestry and documented the potential yields of multiple rotations further influenced private timber companies, especially Weyerhaeuser, to invest in their lands and begin experimenting with tree farms.[37] Starting in 1940, individual states such as Washington also began to heavily subsidize research on tree propagation for commodity production, and state and federal co-operation on fire prevention encouraged private companies to invest in reforestation.

In 1941, Weyerhaeuser established Clemons Tree Farm, the nation's first certified tree plantation. The 130,000-acre plantation was located in a logged, and later burned-over, area in Washington's Grays Harbor County. Weyerhaeuser's decision to replant the area reflected support

for the New Deal's view of trees as crops and forests in need of regeneration. The creation of Clemons Tree Farm also inaugurated the Tree Farm
Program by the National Lumber Manufacturers Association, which
dedicated itself to promoting the idea that forests were renewable crops
and commodities. Given that industry was unwilling to significantly
reform its cutting practices, tree farm campaigns became a savvy way to
appear progressive in the area of forest conservation. In 1937, for example,
Weyerhaeuser hired Roderic Olzendam, a public relations expert, "to
spruce up the firm's image, and the firm began running full-page newspaper advertisements extolling its commitment to resource stewardship
under the slogan 'Timber Is a Crop.'" [38] The National Lumber Manufacturers Association also adopted a resolution that explicitly tethered forest
conservation to the continuous *production* of timber through tree farming,
framing both as "basic to the national welfare." [39]

West Coast timber companies enthusiastically welcomed the campaign
for tree farms. Not only did plantations commit cutover land to ongoing
production of forest crops, but they also provided good publicity to an
industry long criticized for cut-and-run logging. [40] In proactively spread-
ing the message that timber was a crop, the industry both responded to
public concerns and staved off threats of federal regulation in forestry.

Postwar Federal Reforestation
The government's halt on most planting programs during World War
II resulted in a backlog of federal lands in need of reforestation in the
postwar period. The gap between logging and planting continued to
grow between 1930 and 1960. In 1931, the Forest Service estimated
that more than 10 million acres of non-productive forest land (forest
lands with less than 40 percent stocking) existed in the Pacific coast
states. [41] In Oregon, the Tillamook fire of 1933 and subsequent fires in
the same area added another 350,000 acres to the state's understocked
tally. In 1948, Oregon voters passed a constitutional amendment authorizing $12 million in bonds to rehabilitate the land. The first private
contracts to replant the Tillamook Burn were awarded in 1949 and
continued through 1973, when the area was finally declared the
Tillamook Forest. [42] For the most part, though, reforestation was accepted
in theory but not formally implemented or enforced on either private
or federal lands. In a 1960 critique of the Eisenhower administration's
failure to replant public lands, Oregon senator Wayne Morse complained,
"it will take 300 years to do the job for the national forests at the rate
we are going." [43]

Given the successful marketing of trees and forests as renewable crops, why were postwar reforestation efforts stymied, especially on federal land? Historian Paul Hirt argues convincingly that the failure to replant cutover areas had much to do with a lack of political will and, more importantly, financial resources. The case of the Anderson-Mansfield Reforestation and Revegetation Act is telling. While the act expressed Congress's rhetorical commitment to national forest planting, it did not go beyond *authorizing* such funds to actually *appropriating* the necessary amounts. In 1956, for example, Congress approved an insufficient $1.3 million to replant over 4 million acres of cutover forest and reseed another 4 million acres of degraded rangeland.[44] In 1960, Eisenhower's Budget Bureau slashed the original $10 million that Congress had authorized for Forest Service spending on reforestation and stand improvement to a mere $3.2 million.[45] Such budget cuts, which were all too frequent, further hampered reforestation efforts on federal land. By contrast, congressional appropriations for the Forest Service heavily favoured timber sales and road-building activities, programs perceived as generating profit and contributing to economic development.[46] Denied the funds necessary to effectively reforest cutover areas, the Forest Service faced a growing backlog of lands in need of replanting. Thus, although the US government recognized the need for reforestation, it did not provide adequate funds for the effort.

Until the end of World War II, the Forest Service did not engage in active *manual* reforestation of its lands. Most regeneration on federal lands depended on natural methods such as seed dispersal from standing parent trees. These techniques, while inexpensive, did not yield results. In one account, William Greeley highlights the role of "industrious rodents" in foiling seeding attempts:

> Every forest ranger has a tall story of the exploits of squirrels. A ranger in the Bitterroot laboriously gathered bushels of ponderosa pine cones for spring seeding in his nursery. He robbed a few squirrel caches to make up his quota. On returning from a trip to town, he was aghast to find his cone bin all but empty. A little sleuthing showed that a line of gray squirrels were busily moving his hoard back to their own hollow logs in the woods.[47]

There were some cases in which the Forest Service did engage in manual planting prior to and during World War II. Between 1933 and 1942, President Roosevelt's Civilian Conservation Corps operated the majority of its thirteen hundred camps on national forest lands.[48] In addition to replanting degraded areas acquired by the Forest Service through land acquisition programs made possible through statutes like the Weeks Act (1911) and the Clarke-McNary Act (1924), Civilian Conservation Corps members fought forest fires and built campgrounds, fire towers, and trails. Between 1935 and 1942, the Forest Service also administered the Prairie States Forestry Project and supervised the planting of some 217 million trees on more than thirty thousand ranches and farms stretching from North Dakota to Texas.[49] Later, the conscientious objector programs of World War II sporadically contributed workers to the Forest Service's reforestation efforts.

The Forest Service issued its first manual tree-planting contract in the Pacific Northwest on some land near Shelton, Washington, in 1946.[50] By the late 1940s, operations such as extracting *Ribes* (a non-native invasive species of gooseberry), thinning brush, and planting trees were contracted to Anglo operators throughout the region. One of Oregon's first reforestation contractors, Ryan Thomas, recalls working in the woods in the early 1940s as a high school student hired by the Forest Service to help extract *Ribes*. After graduating from high school, Thomas continued to work on public lands, thinning brush and planting trees as a contractor for the Forest Service.[51] Crews in the early 1950s were made up of contractors' acquaintances (often high school buddies) or of men hired out of local employment offices. The prevailing wage was $1.35 an hour – "equal to what a man could make pulling green chain in a sawmill or working as a choker setter."[52] Bob Snow, one of the first reforestation contractors in Washington, notes the high turnover of the workforce in those days: "One or two out of ten would make it, but the ones who did stayed around for several years and gave me a good nucleus to build a full crew around from year to year. We never heard of a workers' comp claim back then ... I think folks were a little bit tougher in those days."[53]

Early contractors had few rules or regulations with which to comply, including labour laws. This lack of regulation also extended to contractor-federal agency relations. Bob Snow recalls:

It used to be that the BLM inspectors would inform us which nursery the seedlings were stored at, we then were required to deliver the trees to the planting sites and they trusted us to do the job right with no current inspection and we did. There was no stashing on my crews. At least if there was I didn't know about it.[54]

During the late 1940s and 1950s, a handful of labour contractors had a monopoly on reforestation contracts. As Bob Snow recollects, the early years were unlike the much more competitive industry that emerged in the 1980s:

> I would like to mention the excellent cooperation we used to have between many of the contractors who competed in the business. If I needed help on a project or needed work myself, I only had to pick up the phone and make several calls to other contractors. I never had to worry about the verbal contracts I made with them. All the contractors I worked with were good about paying and very co-operative.

The Rapid Rise of an Industry

In 1971, Oregon led the nation in passing a stringent Forest Practices Act.[55] Developed in response to growing public concerns about the environment, Oregon's act blazed the trail for environmentally sound forest management standards in a host of states, including Washington and California. The act was groundbreaking in its incorporation of strict environmental quality standards and recognition of forest uses unrelated to timber production, and it ushered in a new era for reforestation. Built upon and considerably revamping the Oregon Forest Conservation Act of 1941 (which only nominally addressed reforestation and failed to define what constituted successful replanting), the new Forest Practices Act formally "required every acre of harvested forestland to be replanted within two years."[56] Reforestation on state, federal, and private lands was thus transformed from simply a recommended activity to a legal obligation, which forced government and private timberland managers to shift their focus to the quantity of trees planted.

The legal requirement to replant cutover areas heralded a booming reforestation industry, as both government and private land managers began to hire large numbers of labour contractors to mobilize reforestation workers. New forest contractors operated similarly to agricultural contractors: they recruited a variety of people, including college-age youth and skid row transients, all of whom were inexperienced, first-time planters. Some contractors narrated stories about driving their "crummies" (vehicles) down the street, grabbing the first dozen people who were willing to work, and heading to the planting unit. The early days of gathering and training crews was largely a "seat of the pants" affair; there were no professional tree planters and people learned on the job.

Labour contracting in the reforestation industry was structured as a top-down chain of command. Contractors, as company owners, hired foremen to supervise crews and liaise with on-site inspectors. Inspectors represented the interests of the landowners and were there to ensure that planting was done according to specified guidelines. On Forest Service land, the contracting officer's representative was usually responsible for inspecting the planting work. If there were problems, he or she informed the crew foreman, who then communicated with the planting crew. Inspectors were only responsible for the technical elements of planting. They were not responsible for health, safety, or labour violations on the crew, leaving contractors and foremen as the only ones responsible for their crews. Thus, the contracting out of work enabled federal agencies to avoid responsibility for injuries or abuse sustained by tree planters and to pass the risks on to private contractors.

The Labour of Reforestation

Commercial tree planting is, and has always been, back-breaking work. A tree-planting job description posted outside the Eugene office of the Oregon State Employment Service read: "It is the hardest physical work known to this office. The most comparative physical requirement is that of a five mile cross-mountain run, daily ... One person in fifty succeeds for the three week training period."[57]

Unlike logging, where technological improvements and the use of machinery have greatly reduced the need for manual labour in the woods, tree planting depends on physical bodies. It is more than likely that reforestation contractors would have adopted mechanized options to replace manual labour if it were feasible. Supervising hand-crews and keeping track of performance takes time in a setting where workers are spread over large areas and are capable of resistance through such practices as shirking on the job and hiding saplings.[58] Because much of the land for replanting in the Pacific Northwest is located on steep terrain, these areas are only accessible by foot. In addition, the best season for planting is during the winter – between December and early March – when it is cold and muddy. As a result, tree planters must haul planting tools and between one hundred and four hundred saplings, weighing forty to fifty pounds, to remote areas that are often covered by slash.[59] Finally, tree saplings are fairly sensitive and must be carefully hand-planted, with their roots pointing straight down.

Through the 1970s and 1980s, a large number of tree planters camped out in the woods, near their planting sites, where they were at the mercy of the elements. Raul Fernández, a former tree planter and crew foreman, remembers being injured in the woods near Klamath Falls, Oregon, in 1972. Like many workers, whether hurt or not, he had no choice but to continue planting and had little access to immediate medical attention or workers' compensation. Fernández recalls the hardships well:

> Once I had a very big job, I started with twenty workers. It was a big responsibility. We all stayed in a camp in the woods. We had no water. We had to bring food and water for the men. Today, it is much easier. Today, they stay in hotels. In those days, there was nothing. We needed to make a camp. We did everything in the camp. I had to attend to everything. And there, we had no bathrooms. To take a bath, we had to fill a tub, boil water with a mountain of firewood and put water and bathe in the tub. We bathed maybe once in two weeks [laughs]. It was not possible to bathe every day.
>
> In mid-October in the hills it was very cold. We put lots of firewood in the heater, and then put the hot water in the tub. I went to bathe. The heater was very hot. I was bathing here [points down] and the heater was here [points down]. As I was getting out, I slipped and fell back on the heater. It was very hot. Bien caliente. I got my mark. It is this big. It cooked my meat. And I was the foreman. I had to do everything. The heater cooked my foot and my back. Oooohhh! Before this, somebody burned up their hand on the heater. Well, the first night, I could not sleep. The guys put cream and a piece of paper towel on my back. And I tried to sleep, but I could not sleep the whole night. The next day, I had to go to work. I was supervising.
>
> And at the end of week, I come to Medford, and showed my boss what I got. "Oh!" he said. "Go to the hospital." But it was already one week. I had already faced the worst part. I got a big scar. And I did not stop. I had to handle all those people on my own.[60]

Organization of the Industry

In those early days, both private landowners and public land managers paid contractors on a per tree basis in order to meet reforestation requirements. This practice incentivized fraud, whereby contractors and tree planters stashed or ditched trees on a massive scale. According to former planters, contractors routinely fired workers if they planted fewer than eight hundred saplings per day. Thus, although contractors might nominally have set rules against ditching trees, their incredibly high daily production quotas, paired with few enforced inspections, enabled an emphasis on quantity over quality planted.

Figure 2.5 The Ridge Runner Timber Services Crew, an Anglo reforestation crew, in the early 1980s. Photograph by Bruce Fraser, published in *ARC Quarterly* (Summer 1983): 7.

To become a contractor in the early 1970s, one only had to have the ability to post performance bonds and have access to a vehicle to transport tree bags to carry saplings, hoedads to plant trees, and a crew of at least six. This relatively low bar of entry into contracting led to a host of unscrupulous entrepreneurs within the industry. Tree-planting crews were often the victims of criminal contractors. Former planters capture the risks of being hired by a scam contractor in a variety of accounts. In one typical case, a crew worked four weeks for an out-of-state contractor. Midway through the contract, the foreman stopped coming to work. The crew completed the planting and asked the contractor for payment. The contractor issued bad cheques and then promptly skipped town. Fortunately, this crew was able to appeal to the Forest Service, which withheld the contractor's final payment until the crew was paid – eight weeks later.[61]

In the late 1960s and early 1970s, most labour contractors were white male US citizens. In this period, a variety of people made up tree-planting crews, including a relatively small percentage of undocumented workers, skid row denizens, mostly male college-age students, and back-to-the-land hippies (see Figures 2.5 and 2.6). By the mid-1970s, the emphasis

Figure 2.6   Another early-1980s Anglo reforestation crew, the Stud City Crew, Evergreen Forest Management. Photograph by Bruce Fraser, published in *ARC Quarterly* (Spring 1982): 16.

on quantity over quality started to change as tree-planting co-operatives emerged and influenced the reforestation industry's standards.

Tree-planting Co-operatives

The Hoedads – the first tree-planting co-operative in the Pacific Northwest – was an innovative and ground-breaking organization in an industry of contractor-run crews. Initially formed in 1971, the Hoedads began as a six-person partnership but grew to one large crew of fifty people by 1972. After a period of rapid expansion, the Hoedads was formally incorporated in 1974, with 135 planters on seven crews.[62]

The Hoedads, and later other co-operative members, represented a fairly well-educated and idealistic segment of the population. Typically, they were in their mid-twenties to mid-thirties, male, and single. If planters were married, it was not uncommon to find both spouses working on crews together. Many co-operative planters had college degrees, and most had middle- and upper-middle-class urban backgrounds. Most important, all planters were US citizens and many had been active in the political and anti-war protests of the 1960s. As a result, co-operative workers brought a sense of entitlement with them. They demanded respect and dignity on the job and took pride in working for a collective good and for themselves. Young planters also brought a great deal of idealism with them. As Hal Hartzell describes it, "There was an intangible feeling, something to do with the 'karma' of planting thousands of trees in the wake of industrial clear-cutting that kept them out there on the units."[63] Tree-planting co-operatives also attracted many experienced planters seeking an opportunity to earn living wages and an alternative to mistreatment by contractors. In 1977, during testimony before a congressional committee investigating reforestation failures, Rick Herson of the Hoedads stated:

> Co-ops and partnerships are stabilizing an industry that does not have a union to protect its workers from exploitation. This greater stability leads to a better understanding of silvicultural techniques and just what it takes to make a tree grow. A planter armed with his knowledge becomes a skilled technician, not just a stoop laborer. I have planted 250,000 trees in 5 years of tree-planting.[64]

Most co-operative workers regarded tree planting as more than simply a job.[65] Many viewed their participation in the Hoedads as a social and political commitment and generally participated in collective meetings and decisions about their organization. Indeed, all but one crew joined the Hoedads for social as well as economic reasons.

Significantly, the one exception to this was Los Broncos, a crew comprising Chicano workers. Although Los Broncos was one of the fastest and most productive crews in the Hoedads, they were also the least engaged in the functioning or politics of the organization. According to Hartzell, "Except for the day they joined, Bronco members never came to Central; they did not come to Council, treasury or bidding meetings. They had not realized the responsibilities required of a co-operative Hoedad crew."[66] Los Broncos members spoke little or no English and were likely unable to actively participate in the daily management of the co-operative. They were also the only non-white members of the Hoedads.

The non-participation of Los Broncos hints at variations in the sense of entitlement and political clout of different co-operative members, probably based on differences in race, class, language ability, and citizenship. Although Hartzell identifies Los Broncos crew members as "Chicanos," it is unclear whether some members were actually undocumented workers or at least lacked US citizenship. Such factors might well explain Los Broncos members' lack of active participation in the co-operative. Their non-participation also points to a lack of racial integration among the Hoedad crews and a paucity of outreach attempts by leaders in the co-operative.

The co-operative structure of the Hoedads offered an alternative to employee-employer relations, and as owners of their own labour, co-op members posed a subtle yet important challenge to class relations. Because they were not traditional employees, Hoedads were initially able to avoid paying into workers' compensation and other employer-related expenses. The Hoedads were thus able to lower their bidding price on competitive contracts, largely through labour self-exploitation. As Hartzell elaborates: "On government contracts a minimum wage was established for employees of contractors, but Hoedads' workers were also Hoedads' owners, not employees. They weren't required to pay minimum wage. In short, Hoedads were able to compete in the market and win work that they needed to keep growing."[67]

Given their middle-class status, educational background, and social ideals, most co-operative members willingly *chose* to work in the woods, rather than being trapped in these jobs out of economic necessity. Thus, the co-operative structure of labour relations and the class background of the co-operative's members probably undermined the minimum wage for everyone else in the industry – especially for those who had no choice about working in the woods. The self-exploitation of co-operative members therefore exacerbated the labour exploitation of low-wage workers by contractors.

The Hoedads also brought about a shift in the rules of the planting game itself. One of the formative debates about work ethics for the Hoedads – and one that distinguished them from contractor-run tree-planting crews – was their challenging the common industry practice of stashing trees. Most early co-operative members, whether experienced planters or not, had joined the Hoedads with a greater social and environmental vision in mind, a vision that also translated into how they chose to make a living. Although members were initially split about whether or not to ditch trees, a majority of Hoedads reaffirmed the decision not to stash trees in the field. Most co-operative members felt that the practice of stashing saplings was dishonest and tainted the relationship a person had with his or her livelihood. For many, working the land was about more than a mere wage; it was also about the experience of learning together. A new precedent was thus established in the industry, one that emphasized quality planting over quantity planting.

Even Forest Service contracting officers noticed a distinction between the social values and work ethic of traditional contract crews and independent, free-thinking co-operative crews. Larry Gangle, a contracting officer's representative in the Alsea District in Oregon, remarked:

Independence is not a benefit in terms of administration. In pure simple facts, it's easier to inspect and keep track of a regimented crew that's operating on a military-like string, than it is a group of individuals who want to do their own thing. It is easier to inspect the old kind of crew, but I don't think you got better work from them. Despite all the problems that the co-ops have, they do plant better trees. Usually they are more interested and want the trees to live. This makes a better quality planter.[68]

By the mid-1970s, landowners and land managers, and even the Forest Service, began switching from paying by the tree to paying by the acre. The Hoedads' focus on quality tree planting partially contributed to the shift away from the number of trees planted and toward a more holistic view of landscape health. Co-operative workers' demands for dignity and respect on the job also underlay the need to broaden the industry's focus beyond the pace of planting. During the same period, Forest Service officials introduced silvicultural improvements that reduced sapling mortality and implemented more rigorous inspection protocols that emphasized planting techniques in addition to mere quantity planted.

The emphasis on quality planting helped contribute to the Hoedads' early success in the reforestation industry. Paying per acre was advantageous to the co-operative for several reasons. First, the Hoedads had already established an organizational culture of quality planting. Second, every co-operative member could also double as a supervisor in the field. With everyone capable of monitoring each other's work, quality planting was more likely. This contrasts with the situation on a contractor-run crew, in which only the foreman is responsible for supervising employees, giving individual employees no incentive to supervise one another. Third, the Hoedads had strong individual crew identities. As members stayed on crews for sustained periods of time, they developed both crew loyalty and pride in their work. Such factors helped contribute to efficient planting and cost-effective intra-crew monitoring and supervision.[69]

By 1975, the Hoedads had established a fairly solid reputation, especially with the Forest Service, and more than 250 Hoedads had participated in the completion of 32 contracts totaling $500,000 altogether. By 1978, 1,000 Hoedads had participated in completing a total of 225 contracts for $6,000,000.[70] The Hoedads' work increased every year, "reaching a peak in 1978, but remaining high in 1979 and 1980. Wage returns increased every year, and by 1980, top co-operative forest workers were earning $25 an hour and $25,000 a year and more (in 1980

Figure 2.7

Membership in tree-planting co-operatives, 1972-89

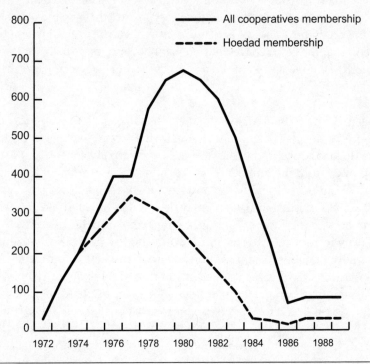

Originally appeared in Gerald Mackie, *The Rise and Fall of the Forest Workers' Cooperatives of the Pacific Northwest* (unpublished manuscript, University of Oregon, Political Science, 1990).

dollars)."[71] With the significant growth and success of the Hoedads, other co-operative efforts also took hold in the reforestation industry.[72] Co-operative membership, however, began to decline after peaking in 1980 (see Figure 2.7). There are a number of reasons for this, ranging from internal problems with Hoedads' rules of governance to significant and effective external organizing to defeat the threat of co-operatives in the industry.

The Decline of Co-operatives and ARC

The Hoedads' rules governing membership and participation eventually contributed to their decline. All co-operative members, regardless of years of membership, had equal ownership and voting power. As the Hoedads grew rapidly (see Figure 2.7), new recruits came to outnumber senior members, and decisions were influenced by the more short-term views of newcomers. Policies around refund periods, access to capital, and other financial matters thus suffered as short-term decisions prevailed. Scholar and former Hoedad Gerry Mackie notes that

> any proposal to reduce the influence of short-term members could be defeated by short-term members. Conversely, those with long-term interests would fail in enacting long-range policies, and, thwarted, would tend to exit the cooperative, frustrating accumulation of a long-term majority. Such a trap can only be avoided at the constitutional stage, through provisions such as a one-year probation before admission to membership.[73]

A number of other factors also radically altered the playing field for co-operatives. The Hoedads and other co-operatives were successful in bidding competitively in part because they were not subject to workers' compensation premiums or minimum-wage laws. To challenge tree-planting co-operatives' exemptions from labour law, contractors banded together in 1974 to form the Associated Reforestation Contractors (ARC), with the stated purpose of "trying to establish a level playing field for competition in contracting."[74] Although few in number, the original ARC contractors were determined to counter the threat of co-operatives and the imposition of rules and regulations within the reforestation industry. ARC officers especially viewed the co-operatives' exemption from payroll taxes and minimum wage requirements as creating an uneven playing field. They mobilized in force by writing letters of complaint to federal and state agencies, taking the matter to court, and starting a quarterly publication to keep industry participants abreast of issues affecting reforestation contractors.

In the 1970s, the federal government required a prevailing wage that was significantly higher than what reforestation contractors were paying on projects for private timber companies. As one of the charter members of ARC noted,

> Payroll taxes during the period when the association formed ran about 40 to 50 percent. Some ARC members felt that they were losing bids on federal contracts to the contractors not paying payroll taxes, which they felt had a competitive advantage. The ARC sued several of those not paying payroll taxes in state court and the courts found those avoiding payroll taxes liable for such.[75]

The "contractors" whom ARC sued were, in fact, workers' co-operatives – in particular, the Hoedads. Between 1974 and 1985, ARC successfully lobbied both the Oregon state legislature and the federal government for uniform enforcement of payroll tax regulations and workers' compensation. As a result of their litigation, the Hoedads and other co-operatives eventually became subject to payroll taxes.[76] Many members of ARC framed their demand for subjecting co-operatives to labour laws in terms of worker exploitation. For example, one contractor wrote in the *ARC Quarterly*:

> Years of seemingly low (at least to me) bid prices by Hoedads combined with reports of low productivity by their crews leave many contractors, myself included, amazed. Given these factors I am not surprised to hear of complaints from Hoedads workers. Neither do I tend to disbelieve Hoedad workers in it be accorded at least the minimum wage standards and benefits as provided by law. Professional standards cannot be maintained when workers are asked to accept below standard wages due to bad bids and poor management.[77]

Ironically, ARC's mission statement reveals a deep concern over growing state and federal regulations that protect or benefit labour, including increases in the minimum wage. The ARC mission statement reads as follows: "Members of ARC are concerned about growing industry regulations, workers' compensation, safety, regional and seasonal volume of available work and rapid increases in illegal alien labor, as well as other state and national issues of importance to our industry."[78]

In one of ARC's successful campaigns, for example, the organization badgered the Department of Labor into rescinding its $10.76 minimum wage dictate on federal contracts in 1985. Previously, the minimum wage had been $8 per hour, with fringes, on federal contracts. According to many contractors, raising the wage by 31 percent in one year would prove devastating to competitive bidding on federal contracts and would lead to "going broke under-bidding government work and underpaying one's employees."[79]

ARC also took on what it perceived as problems with the state's workers' compensation program.[80] As small business owners, many contractors felt that workers' compensation insurance was too high and that employees might take advantage of such benefits to falsely claim on-the-job injuries. For example, ARC publications all have a "Safety Section" that discusses safety practices and "false" cases of workers' compensation claims. One contractor sarcastically commented on the case of an injured worker in Oregon, referring to him as an "Instant Winner. Other workers could not help but be impressed at the length of the paid vacation being enjoyed by this apparently healthy man."[81] Moreover, contractors in ARC wanted to develop "a legal application form that our firms can use to help screen out workers with a greater than average potential to become injured on the job."[82] Of course, ARC publications also consistently endorsed safety training and equipment for crews, as this helped prevent on-the-job injuries and related claims for compensation.

Rather than pushing for compliance with labour regulations, ARC primarily lobbied for equal enforcement so that co-operatives would also be subject to these laws and their related expenses. Consequently, ARC did not necessarily lobby to improve conditions for tree planters and other forest workers, but, rather, to make co-operatives less competitive in bids for federal contracts. By the mid 1980s, tree-planting co-operatives in Oregon were all subject to payroll taxes and the prevailing minimum wage. Co-operatives became less competitive, partially because of these laws. ARC's successful lobbying thus played a significant role in re-establishing the domination of contract crews in the reforestation industry. Political economists would argue that small business owners, as capitalists, would naturally try to keep the cost of labour down so that they might increase profits. Many small-scale contractors note, however, that they do not always earn a profit (and if they do, it is enough just to get by) and, in order to bid competitively, they have no choice but to reduce labour costs. Despite the demise of co-operatives, then, the issue of underbidding remains a prevalent concern among competing contractors.

By 1985, ARC had dissolved, having fulfilled its mandate of contesting the threat of tree-planting co-operatives on public lands. Changes to minimum wage statutes contributed most to the decline of planting co-operatives, but internal disorganization and tensions within the Hoedads, and the changing priorities of individual planters (many of whom had begun to settle down and start families) also played a role. Finally, many Anglo contractors, who had already been in the industry for over ten years, began to retire in the early 1980s. Those who remained were hired primarily from a growing pool of Latino workers and foremen. Combined, these developments facilitated the exit of most Anglo forest workers and contractors from tree planting and paved the way for the gradual dominance of Latinos in the industry.

During its existence between 1974 and 1985, ARC constituted a small but effective group of Anglo labour contractors who felt entitled to participate in the political process and who actively lobbied for certain kinds of forest management and labour law. As former ARC president Jim Stauffer noted:

> ARC ... involved itself in the Forest planning process – testifying on proposed wilderness legislation and on Forest Service and BLM management plans. We believe that issues having to do with timber supply are of great importance to the reforestation industry. We have also maintained close ties with forest industry planning and in dealing with threats at the state and local level to private industry's freedom to practice responsible intensive management of private lands.[83]

Stauffer's reference to "intensive forest management on private lands" refers to clear-cutting, which ARC consistently – and not surprisingly, given the industry's dependence on timber extraction to drive the demand for reforestation – endorsed in its publications.

As is evident through its successful legal battles, ARC clearly had access to both political and financial resources, and leveraged these to ultimately benefit reforestation contractors. While not all contractors were part of ARC, the association helped inform the larger community of (mainly Anglo) contractors about the ins and outs of the industry. Stauffer also sought to improve the industry's prevalent image from one of footloose employers and unskilled workers "to sophisticated firms ... with a positive effect on the forest and on the communities where they function."[84]

Organized forest labour contractors were thus articulating linkages between year-round jobs (for contractors), healthy forests, and healthy communities as far back as the early 1980s. As such, ARC represented an interest group that effectively navigated state and federal channels to lobby for changes that benefited all forest labour contractors.

Latino Immigrants in the Reforestation Industry
In comparison to the Hoedads and ARC, both of whom were politically active in debates over forest labour during the 1970s, today's reforestation industry is meek. This shift may be partially attributed to a shift in the reforestation sector during the mid-1980s from a primarily Anglo contractor and labour force to one increasingly composed of immigrants and non-citizens, primarily from Mexico. ARC publications in the early 1980s voiced concerns about the rampant use of undocumented labour and referred to the "illegal alien" problem in terms of unscrupulous contractors underbidding on contracts and marring the reputation of

the industry. On 19 April 1982, Oregon Congressman Jim Weaver called a meeting to assess the status of undocumented labour in the reforestation market. According to ARC reports about the meeting, "Contractors in attendance agreed that alien useage had diminished on the more 'public' ranger districts, but that overall their use is as great as ever ... Low prices for this year's work was not blamed on illegal alien labor [a view taken by ARC], but rather on tightening supply and demand constrictions."[85]

The hearings on undocumented workers in forestry followed close on the heels of the passage of national immigration policy. In March 1982, Representative Romano L. Mazzoli of Kentucky introduced his Immigration Reform and Control Act of 1981 in the House, while Wyoming Senator Alan K. Simpson, chair of the Senate Subcommittee on Immigration, introduced an identical bill in the Senate.[86] The bills made it a crime to knowingly hire undocumented aliens and were meant to penalize employers for doing so. For once, both reforestation cooperatives and private contractors shared a similar perspective and expressed frustration at the Forest Service's lack of enforcement of national immigration policy. Commenting on a high-profile Immigration and Naturalization Service (INS) raid on thirty-seven undocumented tree planters in 1982, Scott Coleman wrote in the *ARC Quarterly*:

Reforestation contractors sometimes doubt USFS resolve to end undocumented alien labor on their lands. Tales such as happened on the Waldport Ranger District of the Siuslaw National Forest this past winter offer no rebuttal ... So there it was!! Two notices from INS to the USFS that illegal aliens were being employed on a government tree planting project. We waited to see Bob Erny invoke USFS Region 6 policy and default Sharipoff. Well, he did, sort of. On February 12, 1982 Bob Erny defaulted Sharipoff. He also terminated his right to proceed. It must be noted that the termination was not for his employing illegal alien labor, but for failure to cure work deficiency.[87]

Coleman's comments express indignation at undocumented immigrants working on federal lands, but perhaps even more so, at the Forest Service for failing to enforce federal immigration and labour laws. These critics accused the Forest Service of both perpetuating the employment of low-wage immigrant workers through its use of low-bid contracts and ignoring labour and immigration violations on federal land. Similar accusations continue to be levelled today.

CREW OF THE MONTH PHOTO

Figure 2.8   "Crew of the Month." Undocumented Latino forest workers being arrested by Immigration and Naturalization Service officer. Siuslaw National Forest, Waldport Ranger District, OR. Source: *Newport News Times* staff photo, reprinted in *ARC Quarterly* (Winter 1982): 19.

The *ARC Quarterly* also published a picture of the detained, and later deported, workers in its "Crew of the Month" section (see Figure 2.8). Typically, the "Crew of the Month" featured contract reforestation crews in recognition of their planting achievements. By contrast, a photograph of undocumented workers shackled and arrested by a law enforcement officer illustrates the disapproval with which many Anglo contractors viewed Latino tree planters. Clearly, there was no apparent concern for the labour conditions of temporary Mexicans, many of whom were joining the reforestation workforce en masse.

The use of undocumented workers can be linked to the presence of agricultural labour networks already established in Oregon, first through the Bracero Program between 1942 and 1947, and then via fruit-picking circuits of migrant workers.[88] The exploitation of undocumented workers was also aided, albeit unintentionally, when the Carter administration ordered the Immigration and Naturalization Service to stop pursuing workers in order to obtain more accurate census counts in 1980.[89] While these factors undoubtedly contributed to the rise of immigrant workers within the reforestation industry, scholars have not adequately explored the institutional and policy mechanisms, social processes, and networks through which Latinos came to dominate the reforestation labour force since the mid-1980s. I now turn to these developments.

3
From Pears to Pines

As you drive north on the I-5 from Yreka and crest the Siskiyou Summit near the Oregon-California border, the descent into the Rogue Valley is both sudden and picturesque. To the east, beyond the lazy grazing cattle on the hillsides, stands the ridge of Grizzly Peak, still charred from the forest fires of 2002. On the scenic route through the valley lie acres of orchards, fallow fields, a few trailer parks, and newly developed subdivisions. The area is still far from being built out – urban visitors might be charmed by the bucolic vistas and think, "What an ideal place to live!" The allure of the Rogue Valley has led to its rapid growth over the past decade and, since 2002, both Ashland and Medford have made *Money Magazine*'s list of top ten places to retire.[1] But not all those who have moved to the Rogue Valley over the past decade have come as wealthy retirees or urban transplants in pursuit of outdoor recreation and the quiet life.

Until the late 1980s and before the boom in real-estate development, the Rogue Valley was covered in fruit orchards and was nationally renowned for its pears, made famous by Harry and David's mail order business in 1934. In her gathering of orchard memories, *Blossoms and Branches*, local author Kay Atwood describes this place nostalgically:

> For almost one hundred years, residents of the Rogue River Valley have lived with the cycles of the orchard year. Gnarled branches silhouetted against the winter sky. Fragrant blossoms of d'Anjou and Bartlett trees burst out in spring. After the pears ripen in late summer, pickers, ladders and lug boxes appear in the orchards to harvest the fruit. Each year a section of west Medford is devoted to handling thousands of boxes of fruit as they are prepared for processing and shipment. For field workers, packers and those who sell supplies to orchards and packing houses,

the continuing cycle provides a livelihood. The rest of us participate indirectly, economically affected by the agricultural base of Jackson County and aesthetically involved with the physical impact of the orchards on the livability of this good place.[2]

"This good place" is also a landscape materially produced through both natural resources and the hard, manual labour of Mexican migrant workers, who have been involved in farm work as far back as 1943.[3] More importantly, these pastoral orchards and fields have been launching points from which Latino immigrants to the Rogue Valley have moved into forestry.

The federal government, and more specifically the Forest Service and Bureau of Land Management, is the single largest employer of undocumented workers in southern Oregon. On federal lands, Mexican immigrants perform such manually intensive activities as brush piling and thinning, fuels reduction, pest control, and reforestation. Pineros face unsafe working conditions, with little recourse in response to labour violations and workplace exploitation. Since 1985, increasing numbers of Latino forest workers have settled in the Rogue Valley, establishing long-term and rapidly growing immigrant communities. By the early 1990s, Latinization had also spread to the ranks of forest labour contractors. Changes in forest management, immigration policy, and other federal programs have all contributed to a labour market largely composed of documented Latino contractors, on the one hand, and undocumented forest workers, on the other.

Latino Immigrants in the Rural United States and Southern Oregon

During the 1980s, a growing number of immigrants moved to formerly non-immigrant destinations within the United States. Over the span of just one decade, these "non-gateway" areas of the country witnessed a sharp rise in their immigrant populations; the Latino population in new settlement states such as Utah, Minnesota, and North Carolina increased by nearly 183 percent by 1990, surpassing similar growth rates in traditional immigrant-receiving states.[4] Data from the 2010 US Census not only confirm these trends, but also reveal that the growth of Latino populations outside urban areas exceeds both metropolitan and non-metropolitan growth rates for all other racial and ethnic groups.[5] Increasingly, Latinos have settled outside the traditional four "immigrant-receiving" border states of Texas, California, Arizona, and New Mexico.[6] For instance, California's share of the Latino population decreased from

34 percent in 1990 to 28 percent in 2010. Meanwhile, the distribution of Mexican residents by state rose in a number of non-gateway areas. Between 1990 and 2010, the Latino population in the United States grew by 126 percent, whereas the Latino population in Oregon almost quadrupled (an increase of almost 300 percent). Most of the Latino growth in Oregon occurred between1990 and 2000, as the population shot up from 113,000 to 275,000 individuals.

Census figures from southern Oregon indicate a similar rate of growth among Latinos. Between 1990 and 2010, Latinos in Jackson County grew dramatically, both in terms of their share of the overall population, from 4 percent to about 11 percent, and in terms of their numbers, from 5,900 residents to over 21,500. In Medford, the largest city in Jackson County, the demographic transformation was even more pronounced, as the number of Latinos grew by nearly 150 percent from 1990 to 2000, while the city's overall growth rate was 34 percent. The Latino population in Medford has continued to grow since 2000, with Latino residents accounting for more than 13 percent of the city's population of 74,900 residents in 2010.

Census data on households also reveals the steady Latinization of the population in the Rogue Valley.[7] In 1990, only 768 households were headed by someone of Hispanic origin or descent. By 2000, this had increased by 160 percent, to 2,016 households, with almost 44 percent of these households being owner occupied. Such fixed investments illustrate a strong trend towards long-term settlement of Latinos in the area.

The growing number of Hispanic children enrolled in Jackson County schools further indicates long-term Latino family settlement in the Rogue Valley. Between 1995 and 2009, for example, Latino enrolment in Jackson County schools nearly tripled. Furthermore, Latino students accounted for 73 percent of the total minority student population in Jackson County and almost 18 percent of all students enrolled in the county (see Figure 3.1). Student demographics in the Rogue Valley are very similar to those in Jackson County as a whole, with Latinos accounting for 17 percent of the overall student population and 71 percent of the minority student population. Overall, census data reveal similar trends in the other towns and cities in southern Oregon and reflect a broader national trend in the growth of Latino immigrants in the United States.[8]

Despite the considerable growth of the Latino population in rural parts of Oregon, the media spotlight has trained primarily on immigrants in industrial occupations who are geographically concentrated in the Midwest or the South.[9] In non-metropolitan areas of these regions, the growth of Latino populations is linked to the rapid consolidation,

Figure 3.1

Hispanic students in Jackson County schools, 1995-2009

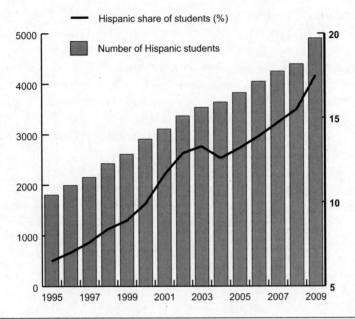

Source: Oregon Department of Education, *October 1 Enrollment by Ethnicity.*
Source: Oregon Department of Education, *Reports: Students* (Salem, OR: Oregon Department of Education, 2010), http://www.ode.state.or.us/data/reports/toc.aspx.

expansion, and restructuring of agro-industrial and manufacturing operations.[10] These jobs involve low-wage non-union workers in the private sector, and most people have heard stories about immigrant exploitation in slaughterhouses and sweatshops.

Yet how many think of labour exploitation on public land? When scholars identify Oregon as an emerging "immigrant destination," they point to the growing Latino population in the Portland metropolitan area and highlight urban job opportunities in the service, retail, and construction sectors.[11] References to Latino immigrants in the *rural* Northwest, by contrast, usually concern longstanding populations of farm workers in areas such as Yakima, Washington, and Woodburn, Oregon, and once again reinforce the focus on labour in the private sector.[12] While work in the private sector is no doubt critical to immigrants, this is only part of the story. In western states like Oregon, where over half of the land is under federal ownership, there are few studies of recent immigrant growth and settlement.[13] It is imperative to consider how employment opportunities on federal lands also shape immigrant growth and settlement.

Agriculture in Oregon and the Rogue Valley

In the twentieth century, the migration of Mexicans and Mexican Americans to the Pacific Northwest was largely motivated by the construction of railroads and the development of irrigated agriculture. Between World War I and the beginning of the Great Depression, between 750,000 and a million Mexicans entered the United States.[14] Migrants to the Northwest found work in sugar beet fields, especially in Idaho, Oregon, and Washington. With the start of the Great Depression and the loss of employment nationally, the stream of Mexican immigrants ebbed. Throughout the 1930s, repatriation programs intimidated, coerced, or led to the voluntary return to Mexico of nearly a million people of Mexican descent. These schemes were aimed at reducing the nation's growing unemployment, but they targeted both Mexican immigrants and US citizens of Mexican descent.[15]

In the mid-1940s, Mexicans again came to Oregon as temporary agricultural workers under the Bracero Program. The Bracero Program, passed as Public Law 45 in 1942, was a response to the wartime labour shortages in the United States. The program was a binational labour agreement between the United States and Mexico whereby access to low-wage Mexican labour was provided to American farmers. By 1943, the importation of Mexican workers had become institutionalized in US agriculture. Of the 220,640 labourers who entered the United States under Public Law 45 between 1943 and 1947, 21 percent were contracted to farmers in the Northwest and over 15,000 were sent to Oregon.[16] Between 1942 and 1947, labour contractors brought Mexican workers into the Rogue Valley to harvest pears in the area's many orchards. The first labour camp in Medford was established in 1941, and by the height of the harvest season in 1945, 400 Mexican workers were living temporarily in Medford.[17] By the winter, with the harvest season drawing to a close, only 180 remained for work maintaining the orchards.[18]

Although temporary in nature, agricultural labour camps faced initial opposition in the Northwest. Local resistance was often based on fears around property devaluation, crime, and pressures on the educational system. Growers, by contrast, feared that the camps would become "hotbeds for farm labour unionization or would bring in hordes of undesirables."[19] The Farm Security Administration held public meetings to quell such fears and exercised its federal mandate to proceed with the construction of camps, which ran under the auspices of the Bracero Program until 1947.[20]

Because the US government limited the number of visas it granted to *braceros* (literally, "one who works with his arms"), thousands of Mexican labourers also entered the country illegally. In 1951, under pressure from agricultural growers, Congress passed Public Law 78 to give the Bracero Program a permanent statutory basis. Mexico excluded Texas from this program on the basis that Texan growers had violated migrants' contracts and civil rights, and, more generally, discriminated against Mexican workers. Texan growers continued to hire Mexicans who illegally crossed the Rio Grande into Texas, although growers did not provide them with transportation, fair wages, decent housing, or health services.[21]

Increasing reports of migrant worker exploitation, coupled with the loss of agricultural jobs for native US workers and white Southwest citizens' perception of "wetbacks" as criminals, eventually led to a national anti-immigrant backlash. In 1954, the Immigration and Naturalization Service (INS) launched Operation Wetback, which primarily targeted illegal workers in the Southwest, starting at the Rio Grande and fanning northward. Between July and October 1954, the INS claimed that over a million undocumented migrants were apprehended and deported, though the actual number of apprehensions did not come close to this total. In addition, numerous illegal migrant workers voluntarily repatriated themselves rather than getting caught in Operation Wetback sweeps. Local INS officials in Texas claimed that between 500,000 and 700,000 workers fled before the start of the campaign.[22] Over the same period, growers repeatedly clamoured to the US government about a national labour shortage and demanded more foreign labour. As a result, the number of bracero visas granted was doubled, reaching 450,000 by 1960.[23] Under the leadership of Cesar Chavez and Dolores Huerta, however, the United Farm Workers raised a public outcry against the slavish importation of foreign workers into the United States and exerted enough political pressure to end the Bracero Program in 1965.[24] Since then, undocumented migrants who remained in the country after the program's termination, supplemented by new arrivals, continue to fill the ranks of agricultural labour in the western United States.[25]

In the post-bracero era, undocumented migrant workers continued to pass through southern Oregon, working the orchard circuit. After the August-to-October pear harvest in the Rogue Valley, many migrants moved on to pick apples in Yakima, Washington, followed by olives in California, and then back to southern Oregon to prune pear trees. Like their former bracero counterparts, undocumented migrant farm workers lived in isolated labour camps. In 1981, approximately fifteen hundred Mexicans migrated through the Rogue Valley and stayed in more than thirty-one camps run by Jackson County orchard owners. These labour camps provided migrant workers with only the most rudimentary shelter and were often cited for health and safety violations such as lack of running water and electricity. Located in or near orchards, these camps were out of sight of the valley's settled residents and clearly delineated the bounds between Medford's white population and the transient agricultural workforce.[26]

While invisible to most residents of the Rogue Valley, migrant labour underpinned the orchard-based economy of the region. Prominent area growers such as Ned Vilas relied on immigrants to get the harvest done and recognized their dependence on foreign labour:

If it wasn't for the Mexicans, we'd never get the fruit picked. They're very good pickers. They're hard workers and when you have a Mexican crew you know you can count on getting so many boxes of pears off every day. They're dependable, and they don't fool around. They don't sit down every fifteen minutes. They'll come down out of a tree at twelve o'clock for ten minutes, eat their sack lunch and go right back on the ladder. They want to make as much as they can.[27]

Although some valley residents acknowledged the growers' dependence on migrants, they also justified, and thus mitigated concerns about, labour exploitation by noting that Oregon agriculture provided jobs for economically desperate Mexicans. One long-time orchardist, A.C. Allen, Jr., described Mexican workers as being "tickled to death" to have work. For Allen, labourers' access to income trumped the exploitation they experienced while on the job:

The reason the Mexicans came in, they were brought in by the government to help harvest the fruit because after welfare and all this socialized stuff came in, natives didn't want to work that hard and fruit had to come off. So they got the Mexicans in here who were used to work, that's what they like, what they wanted to do, at least they did, and they got the fruit off. Then, of course, they stopped that ... When the Mexicans first came in here, of course, they would work for almost anything because it was much more than they got in Mexico, anyway. Undoubtedly they were exploited to some extent. But on the other hand the Mexicans were tickled to death because there was no work in Mexico, so it helped them that much. And they took their money back with them, what they didn't spend here. The merchants profited by it because they were buying all kinds of stuff. Luggage for one thing. You'd see these people buying luggage, mostly cheap luggage. [I] remember, we'd see 'em packing luggage around before they went back to Mexico ... taking their goodies home with them.[28]

Labour camps were located on the outskirts of town, far from Medford's residential and commercial areas, and the comings and goings of the migrant population were never openly discussed by locals. Migrants' social and spatial segregation resulted in, and was a product of, racial anxieties among the Rogue Valley's white population. As one long-time resident of Medford stated:

Medford used to be a lily-white community for so long. It's only natural for people to feel scared about newcomers. They were different. Al Gonzalez was the Mexican farm labour contractor in this area in the 1950s. There would be thousands of workers coming into this area for picking. No one ever talked about them. Everyone just breathed a sigh of relief when the season was over. Al was real good. He just brought these people quietly in and shuffled them quietly out without any problems. Of course, his kids probably had problems. They were the only Mexicans in the school system back then [1960s]. I used to be so scared when my husband went out of town. We used to live next to the pear orchards and when he was gone, I wouldn't be able to sleep the whole night.[29]

Other residents spoke about the "Mexican problem" in relation to immigrants who stayed on in the Rogue Valley. Migrant workers were only unproblematic as long as they remained a source of pliable, temporary, and invisible labour. In the few instances when these same workers chose to settle down and become visible, they were perceived as trouble. Again, long-time orchardist A.C. Allen, Jr., illustrates this pervasive attitude:

Then, of course, the wetback come in here, came in here without permission, and they'd have to round them up. So that's a problem. And I think the Mexican problem is still another impact that has hit this valley which we didn't have. I notice an awful number of the serious car accidents are Mexicans now. There was one last night on Table Rock Road on the overpass over I-5. And one was driving a pickup with several people in it, skidded, and rammed into another car, nine people were injured, I guess none of them fatally. But it was a strictly Mexican name, Rodriguez, or something like this.[30]

Allen's account repeatedly focuses on the "illegal" and problematic behaviour of immigrants in the Rogue Valley. Whether it was getting into car accidents or creating other trouble, for Allen and others with a similar perspective, Mexicans were always the ones perceived as being at fault.

Settling in the Rogue Valley

What first compelled Latinos to *settle* in rural southern Oregon, an area in which they had few social ties? Until the late 1980s, the majority of Latinos in the Rogue Valley were passing through on the migratory fruit-harvesting circuit. Still, a few Latino families did settle in and around Medford in the early 1970s. These early settlers were newcomers to Oregon, but not to the United States. They moved to southern Oregon with wives and young children, and they all had prior work experience in agriculture or on cattle ranches in California. Some of these settlers first arrived in the United States as braceros and were later sponsored for legal residency by their employers in the wake of the Bracero Program's termination in 1964. Others crossed the border without papers. They were introduced to life and work in the United States by more experienced male relatives and decided to stay in the country.

The availability of forest work, coupled with the need to support young families, played a critical role in migrant workers' decisions to stay in the Rogue Valley during the 1970s. María López, for example, moved to Medford in 1972 when she was still an infant. Her family was among the first Latinos to settle in the area. Like other early immigrants, María connects her life in the United States to the labour of her grandfather and father. Her grandfather, the first of her family to make the trip to *El Norte*, was a bracero during World War II. He never settled permanently in the United States, but his sons did. María's father and uncles, however, were not braceros. They all crossed the border as undocumented workers. As María recalls:

My grandfather was telling us the story of his life and how it all started. He said that while the war, World War II, was happening, they didn't have enough people to come and work so the US government decided that it would go to Mexico and contract people out for a period of six months. And my grandfather was one of the people who was contracted ... "braceros," they called them. They brought him to work in Ohio to do train tracks for the US Army so that they could get the mail through. He said that when they came, no one spoke Spanish. So they had a really hard time communicating with the soldiers and everyone else. He's got some wild stories. But he doesn't speak English.

Grandfather never lived in the US permanently. His kids are the ones that started coming after he introduced them to the US. He brought my dad and his two older sons to work in the US for the very first time. They started working on cattle ranches in Crescent City, in northern California.[31]

Father was not a bracero. The program ended. Grandfather went back to Mexico but the Revolution was going on in Mexico and he couldn't make it. He decided to come to work in the US whenever he could. But he never brought his wife or the other children. There were thirteen children. So he started bringing the three oldest first. And then they decided to stay. Like when my parents came on their honeymoon, they stayed. I mean that was just it. They just decided to come ... Dad brought my mom to a ranch where he was already working. In Crescent City. Mom was pregnant with me about a year after they got married. It was only later that they got their papers.[32]

Pedro Sánchez, like María López, was among the first Latinos to settle in the Rogue Valley. A contemporary of María's grandfather, Sánchez first came to the United States on a bracero visa. He remembers the lonely years he spent working in New Mexico and moving from one isolated ranch to another, before having enough money to bring his wife to the United States. His story not only highlights his years of work experience in the southwestern United States, but also illustrates how he fostered relationships with his American employers so that he was able to successfully find a sponsor for his Green Card. For María López and Pedro Sánchez, crossing the border illegally was relatively easy prior to the 1980s, and it was possible to obtain legal status for family members once they had arrived in the United States. Sánchez remembers:

My story: I came as a bracero in 1957. In September 1957. I worked as a bracero for three years on the mesa of New Mexico near Las Cruces. From there, I went to Saffor, Arizona. I worked in the field. Then I went to an Indian reservation in Arizona. And there I worked as a cowboy. I herded horses. A very big Indian reservation in San Carlos, Arizona. And then, I worked in another ranch for eight years as a cowboy. When I was a bracero, I earned fifty cents an hour. Then, later, I started driving a tractor, I earned sixty cents an hour. I not married then. I was single. I am too old right now. But at this time, in 1960, I was thirty years old ... When I came to Arizona, I began at $150 a month. When I started being a bronco rider, I started getting $180 *por mes*. There were very few people there. I was alone there, in that place.

After three years as a bracero, I arranged my papers. I then went to Mexico for a week, and my boss hired a lawyer. And in one week I came back. I entered with papers, in one week, in 1960.[33]

As some of the first non-white immigrants to settle in the Rogue Valley in the 1970s, numerous Latinos and Latinas experienced social marginalization, physical segregation, and racial antipathy. Often, it was the children of newly arrived immigrants who most poignantly recalled instances of social hostility. Alejandra Puentes, who came to the valley when she was seven years old, reflects on those times:

I remember one time we had just gotten here and we had an old '67 Mustang. And I remember my dad had it like jacked up in the back. And the family was together and we were driving down Riverside [Avenue]. And there was a little old man who wanted to cross the street. So my dad stopped and let him cross. And when he got to the island in the middle of the street, he stopped. I'll never forget this – it was an impact, as we had just gotten here, you know. He stopped in front of our car and with his cane, he started showing it to us and said, "Go back where you came from. Leave us alone, you're not wanted here."[34]

The humiliation was public, and it hurt. Alejandra's father was shunned from the central street of downtown Medford and metaphorically cast from the city's core. As new arrivals seeking to make the Rogue Valley home, the Puentes family was told that their presence in the heart of the city was unwelcome. The old man with the cane was engaged in policing the boundaries of where Latinos could or could not be. Immigrants were expected to leave after the fruit season was over, and Alejandra's father had breached many Anglo residents' unvoiced expectations about Latinos' mobility.

At the crossroads of immigrant and non-immigrant worlds, Latino children also endured daily practices of exclusion, often perpetrated by authority figures, especially in spaces like the classroom. With so few non-white youth in Rogue Valley schools during the 1970s, Latino students became easy targets for derision. Lucía Tomás was born to undocumented farm workers in Fresno, California. In her early years, she was part of a tight-knit extended family and was surrounded by other Latinos. At the age of eight, Lucía and her family moved to the Rogue Valley so that her father could find work in the woods. For the first time in her life, Lucía was the only Latina in her classroom:

> I've lots of experiences. I was called a lot of bad words all the time. Every day. I fought a lot in grade school. I remember that some kid asked a teacher why we had to go to school. This is third grade. "Teacher, why do we have to go to school?" And he looked out the window, and there was an orchard at the front of the school and at the side of the school. There is a park there now. And he [the teacher] stood us all up and said, "Look out the windows. You have to go to school so you are not like those Mexicans picking pears." And I got up and I said, "My dad is a Mexican and he doesn't pick pears. And those people are here because they need to feed their families, not because they want to be [there]. I'm not embarrassed of what they are." And I got in trouble for talking back to the teacher.[35]

Another child of pineros, Lydia Rincón was born in Medford, Oregon, and is a US citizen. As a child of Latino immigrants, she was nonetheless cast as an outsider by school authorities and labelled as "non-American." Lydia recalls:

> My most memorable experience at Orchard Hill was that my first grade teacher didn't let me say the Pledge of Allegiance because I was not American. Or because my parents were not American. At that time, I didn't think anything of it. I was only in first grade. I thought, okay, I'll just go stand outside. I didn't really realize that something was wrong until a substitute came into the class and one of the students had to remind me that I had to go outside. And the substitute was like "Why?" And I had to go, "Well, doesn't he know that I'm not American?" And then it started clicking that maybe it wasn't right. I was American. But ... my teacher told me otherwise. So of course I'm going to believe her. I didn't tell my parents about it because I didn't think it was a big deal. It wasn't until high school that I started thinking about it more.[36]

For most immigrant children, life in the Rogue Valley was not carefree. Fluent in both English and Spanish, these youngsters had to interpret for their monolingual parents and navigate within various arenas of the public sphere. Both insiders and outsiders, yet never completely part of either world, these children felt particularly isolated. Lupe Ramos, who was also born and raised in Medford, recollects her daily discomfort in public spaces:

> It was hard. There was a lot of racism. There still is. But not as much anymore. Ever since Taco Bell introduced their burritos, it kind of introduced our culture. But there was a lot of racism. I remember eating a burrito and people in school would say, "What's that?" I went to school and learned English. People didn't know because we were very few, we were criticized. We couldn't go anywhere because there were not any places that were bilingual. You would go to the store and no cashier was bilingual. You would go to the public agencies and there was not one person that spoke Spanish.[37]

Such experiences of social exclusion from the Anglo population touch on the all-pervasive reality of Latinos' restricted mobility during the 1970s and 1980s. When Alejandra's family was told to "go back where you came from," they were labelled as intruders to the Rogue Valley and publicly insulted. Similarly, Lucía's experience in school highlights how even respected authority figures like teachers reinforced perceptions of Latinos as belonging only in the orchards, as agricultural labour. Lydia's daily expulsion from the classroom during the recitation of the Pledge of Allegiance and the restricted mobility of Lupe's family due to their poor English also reveal the extent to which most Latinos, regardless of whether or not they were US citizens, were perceived by white residents as neither "American" nor as belonging to the Rogue Valley.

Immigrants' experiences of social marginalization, however, did not prevent them from establishing a sense of place or belonging in the Rogue Valley. Immigrants congregated at the Catholic church in particular, a place of sanctuary for undocumented workers to apply for legal status under the 1986 Immigration and Reform Control Act. In Medford, the Sacred Heart Catholic Church's pro-immigrant stance may largely be credited to the dedication and involvement of one individual, Sister Yolanda. Sister Yolanda arrived in the Rogue Valley in the early 1980s, when the increase in the Latino population was still quite recent. She spoke Spanish and reached out to the Latino community. The church began offering a Spanish language mass and organized youth groups. Ester Moraga, who grew up attending Sacred Heart Catholic Church, remembers:

I feel like that Sacred Heart Catholic Church has been really strong. They were the one site where people could come and fill out their amnesty forms. So we saw a lot of people ... As the families started arriving we started doing more things. The church was where we used to see each other on the weekends. In our culture, your main community is who you go to church with.[38]

Today, most immigrants in the Rogue Valley are no longer socially isolated. They enjoy the services of numerous businesses and social and cultural organizations that cater to the Latino population, including Latino/Hispanic grocery stores, eateries, bars, money-changing services, churches (both Catholic and a growing number of evangelical congregations), soccer leagues, community health centres, English as a Second Language classes, a Latino Chamber of Commerce, and the Hispanic Interagency Committee. As the Latino population has grown, some lament the loss of a once tight-knit immigrant community. One long-time Latino resident notes:

Medford is packed with people. You see people that you had no idea existed. Talking back to when we first came, there was no mass in Spanish. I compare back then to now a lot. We now have two bilingual, Hispanic masses at Sacred Heart on Sundays. I think the capacity of the church is six hundred. There are always people standing in the back. From a community when there was three families that knew each other, you don't know anyone anymore.[39]

As more Latinos settle in the area, it is likely that social and cultural organizations will continue to grow as well. Today, most organizations serving Latinos are primarily service or culturally oriented, and meet basic social needs. They do not, however, represent any significant political power on the part of Latinos in the area. Although the Latino population has grown tremendously over the past decade, they have not been politically active, nor have they been included in policy decisions that affect their daily lives. The political invisibility of Rogue Valley's Latinos also extends to the workplace. Even though Latino forest workers and contractors are "dominant" in forest work in the area, they often have little say in highly publicized forest management debates, and lack access to information about certain types of forest procurement contracts, and forest workers have a limited ability to change poor working conditions or confront labour violations.

The Latinization of Forest Work

Most Latino families who first came to the Rogue Valley settled in the area because they were able to find stable work in local sawmills or as tree planters on federal lands. Gustavo Flores's story is typical of many Latinos' entry into forest-related work. His father found a job in the local mill and his entire family moved to Medford on a week's notice:

> My parents and I lived in Los Angeles, but my dad was tired of the lifestyle in LA. So we came to Medford to visit my uncle [who worked in the tree-planting business] in August. Came on a Thursday and dad was working at a mill on Saturday. And he sent us back to get our things and that was it. We came to visit and ended up staying.[40]

Others, such as Braulio Santos, were directly recruited to the woods while already working in the fields. Braulio's story highlights how social relationships between early forest labour contractors and orchard supervisors (foremen) facilitated the transfer and use of Latino workers from private fields and orchards to public lands:

> I came to Oregon to pick pears for two months in 1972. I stayed in the company house with three other families. Some pineros came to the orchards to borrow some workers. Al Gonzalez was friends with the foreman in the orchard so the foreman let him, Al, take some workers to the woods. I then went back to Modesto but didn't like field work ... I came back to Oregon in 1976. I found work in Medco mill and put laminate on press boards. I was at the mill for fourteen years. Most years, I was laid off from the mill for one or two months. In this time off, I went to plant trees. In a week, I became a foreman.[41]

Braulio's promotion to foreman also hints at the gradual rise of Latino workers within the reforestation industry. Although Anglo contractors ran the industry in the 1970s and early 1980s, they depended heavily on Latinos to supervise crews and recruit workers through immigrant social networks. Thus, when Anglos began to retire from contracting in the mid-1980s, it was the more experienced pineros who moved to fill their ranks.

Latino Forest Contractors

In 1980, 106 forest labour contractors were licensed in Oregon, but in the 1970s, Jim Daniels' Rogue Valley Reforestation was the major player in southern Oregon.[42] Daniels' company served as a springboard for a few pineros to move into foremen positions. In 1972, Rubén Gómez became Rogue Valley Reforestation's first Latino foreman. Gómez went on to play a central role in the early Latinization of the area's forest workforce by recruiting his male relatives and friends into the company's reforestation crews. Gómez hints at the larger demographic shift in the workforce – from Anglo to Latino workers – and highlights how being a foreman allowed him to accumulate the work experience, capital, and social networks necessary to eventually become a contractor:

> In 1969, I came to Medford with another person to pick pears in mid-August. People would say there is a place to work planting trees. From here, after we finished picking pears, we went to Washington to pick apples. When I returned from Washington in November, I started to work for a private company, planting pines. The company was a private forestry company. The company was called Rogue Valley Reforestation. They don't exist anymore ... I worked with this company for eighteen years. Straight. When I quit this company, I making my own company. So in this work, I worked three years as a peon. In three years, my employer promoted me to *mayordomo* [foreman]. It was a very big appointment. When I began my first job as mayordomo, I had to plant 1 million trees with twenty men. It was a very hard job. We couldn't plant them all because of the snow. We planted 600,000. I had to manage twenty men ... When I start collecting this money, I had my own money to start my company.
>
> When I started working with the company, there were many Americans. Many, many good workers. Even women. I think ten or twelve women worked with us. Good workers, good workers. The work was ... A few guys ahead are scalping the trees and drilling holes and the women come and put the tree in the hole and cover it up.[43] One time, I drilled 6,600 holes. In one day. And one American woman and one Mexican man behind me, they planted the trees. Pay was one cent a hole for me. And pay for the woman I think was two cents a tree.
>
> There was another foreman in the same American company who brought some hippies. He brought these hippies and they stayed until about 1975. After that, all the Americans left. And then it was purely Mexicans. When the Americans left, I was left with only Mexicans.[44]

In 1988, after working for Jim Daniels for eighteen years, Gómez became the first Latino reforestation contractor in the Rogue Valley. A handful of other pineros, who had all been hired by or worked under the supervision of Gómez, followed suit in 1989. In order to become licensed labour contractors, these new Latino entrants had to meet basic requirements, including US citizenship or permanent residence, sufficient capital to start a business, experience in forest work, knowledge of how to bid on federal contracts, the ability to keep records, and access to labour. The majority of these Latino contractors had received their permanent residence through the 1986 Immigration Reform and Control Act, which granted amnesty to those who were in the United States prior to 1982.

Immigrant family networks also played a significant role in providing access to the financial capital and labour necessary to become contractors. In 1989, for example, Gómez's nephew Juan started his own reforestation company with help from his already established uncle. Juan's labour crew consisted mainly of family members, including his brothers and several nephews. Two other contractors were also close relations: Pedro Martinez ran one company with his first cousin, Ricardo Sánchez, and Pedro's daughter, Emma, ran another company with her husband, Jesus Rodriguez. All of these contractors asked relatives for financial support and, in many cases, expected family and friends to contribute labour without compensation or with deferred compensation. While such labour practices are illegal, they reveal how critical family networks were in establishing the presence of Latino contractors.

Male relatives were on the front lines of unpaid forest work, but women played an important supporting role. When Juan started his company, his mother, Señora Guadalupe, subsidized start-up costs by making thousands of bags for the saplings, which was much cheaper than buying the materials retail. Señora Guadalupe also sewed and supplied bags to other contractors, giving her full earnings to Juan's business. Other family members, both male and female, helped keep Juan's costs down by contributing their skills in repairing vans, welding, and carpentry. Finally, Juan's female relatives took the lead in rearing livestock and providing their extended family with homegrown produce. When I first met Señora Guadalupe, she took me on a tour of her back yard. Amidst the rundown automobiles, rusty chainsaws, and parked vans, thirty-six hens and two cocks ran about foraging for scraps: the family never buys eggs. In an adjoining half-acre lot, Señora was rearing three large cows, one of which was slated for butchering later in the month. Because the family grew much of its own food, Juan was able to save on household expenses and reinvest his earnings into his growing reforestation business. Thus, while women may not have been wielding chainsaws in the woods, they nevertheless sustained the labour of male relatives who did so.

Figure 3.2

Latino forest contractors in the Rogue Valley, 1980-2005

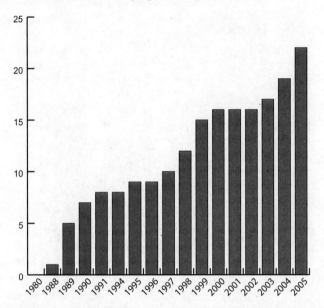

Sources: Oregon Bureau of Labor and Industries, *State of Oregon: Farm Labor* (Salem, OR: Oregon Bureau of Labor and Industries, 2010), http://www.oregon.gov/BOLI/WHD/FFL/index.shtml; General Services Administration, "Central Contractor Registration," 2010, https://www.bpn. gov/CCRSearch/Search.aspx; interview data.

Although there were no Latino contractors in the Rogue Valley prior to 1988, five Latino-run firms emerged in the span of only two years (see Figure 3.2). After an initial levelling off from this spurt, the number of immigrant contractors grew steadily through the 1990s. By 2000, Latinos headed sixteen contracting firms. As the number of Latino firms grew, so too did their share of the overall contractor population, from 38 percent in 1990 to 76 percent in 2000 (see Figure 3.3). This rapid Latinization of the contractor population parallels the disproportionate Latinization of the forest labour force as well. While there are no publicly available figures on the ethnic composition of the workforce, interviews with forest workers, labour contractors, Forest Service officials, and Bureau of Labor and Industries inspectors indicate that Latinos make up the vast majority of the forest management workforce today.[45]

Latino contractors have become significant players in forest management, not only because of their growing numbers, but also because they are getting some of the largest contracts awarded by the federal

Figure 3.3

Forest contractors in the Rogue Valley by ethnicity

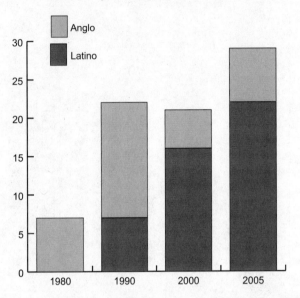

Sources: Oregon Bureau of Labor and Industries, *State of Oregon: Farm Labor* (Salem, OR: Oregon Bureau of Labor and Industries, 2010), http://www.oregon.gov/BOLI/WHD/FFL/index.shtml; General Services Administration, "Central Contractor Registration," 2010, https://www.bpn. gov/CCRSearch/Search.aspx; interview data.

Figure 3.4

Latino and Anglo shares of Forest Service contracts, 1999

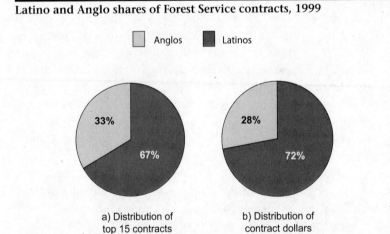

Source: Ecosystem Workforce Program, "Data Provided by Ecosystem Workforce Program on 1998 and 1999 Total Market Share of Region 6 Forest Service Awarded Forest Management Contracts" (University of Oregon, 2004).

government. For instance, in 1998 and 1999, the Region 6 Forest Service awarded approximately $25 million worth of contracts to forty-five forest contractors in Oregon.[46] Twenty-seven of these contracts went to Latinos.[47] Latino contractors also won the five largest contracts (in terms of dollar amount) and were ten of the top fifteen awardees. Three of the top five awardees were Latinos from the Rogue Valley. Overall, Latino contractors made up 60 percent of the total population of awardees and received $17,421,810 (or 72 percent) of the total dollar amount of contracts (see Figure 3.4).

Today, the average number of workers per contractor ranges from forty to eighty-five employees, with contractors running from two to eight crews each. Contractors engage in a number of different types of work on federal lands, including tree planting, pre-commercial thinning, brush and slash piling, slash burning, scalping, tubing, mulching, cone collection, site preparation, hand weeding, gopher baiting, trail maintenance, nursery services, and fire fighting (see Figures 3.5 and 3.6). The variety of tasks performed by labour crews not only reflects the manually intensive nature of forest work, but also highlights the shift in land management priorities from reforestation (tree planting), which was related to timber extraction, to more ecosystem restoration-oriented activities such as fuels reduction.

The growth of Latino contractors has not necessarily meant progress for pineros. The increasing success of Latino contractors has largely depended on the perpetuation of a segmented workforce based primarily on divisions by legal status. Many workers note that in return for employment, they are expected to lend foremen or contractors part of their earnings and work longer hours without pay. Relatives and others in contractors' social networks are dependent on or indebted to documented contractors and foremen, who have financially helped to sponsor their journeys, whether legal or illegal, from home states in Mexico. The story of Alberto Morales is a case in point. Alberto began planting trees in 1983, first working for his uncle and later as a foreman, under his brother. Alberto was responsible for supervising workers and transporting crews and a trailer full of equipment to the planting site. On one of these trips, Alberto suffered a debilitating back injury while trying to dislodge the trailer, which was stuck in mud. Alberto was unable to work for three months and received no workers' compensation. When I asked why he did not report the injury, he replied: "Do you know how to tell that a bucket full of *cangrejos* [crabs] are Mexican cangrejos? When one cangrejo tries to climb out of the bucket, the others keep pulling him back. That is the way it is with us; our own people keep us

Figure 3.5   Latino forest worker picking Douglas fir cones. Provolt Seed Orchard, Oregon (Bureau of Land Management land). Photograph by Brinda Sarathy, 2004.

down ... Here it is Mexicans who report others to *la migra* [immigration authorities].[48]

As a result of his injury and subsequent unemployment, Alberto's wife, María, stepped in to transport workers and to help clear brush and remove branches in the woods. María noted that it was not uncommon for the wives of pineros to assist unofficially in forest work. She added that many women also find employment in Medford's pear-packing

Figure 3.6   Latino workers hand-weeding sapling beds at Stone Forest Service Nursery, Medford, OR. Photograph by Brinda Sarathy, 2004.

plants to tide their families over during periods when their husbands are waiting for work. Latinas faced various kinds of abuse at work in order to make ends meet. María remembers her own time packing pears at Bear Creek:

> I worked at Bear Creek for six years. I boxed dry fruits. I been out of work for six months now. Don't have insurance. And I have type 2 diabetes. It is very hard for the women when the men [pineros] get hurt in the woods. They stay at home and drink. They have no insurance. I tried to organize a work stoppage at Bear Creek. The problem was that the Mexican women would pack twenty to twenty-two boxes a day and the Anglo women could only pack eighteen. At work, the Anglos ignore you. And other Mexicans can report you to *la migra*. Another Hispana [Latina] worker reported me [to management] when I tried to organize the stoppage.[49]

The disparity in legal status between documented employers and undocumented employees often leaves undocumented workers reliant upon social networks to find work. The undocumented status of many pineros renders them highly vulnerable to threats of deportation and job loss, and most avoid contacting federal officials to report workplace injustice. As a result, although many of the pineros I interviewed had experienced labour violations or sustained injuries on the job, none had reported their contractors to the authorities or claimed workers' compensation. Moreover, it is not only pineros who are affected when they are injured or unemployed. Their wives often help out in forest work and take up other employment to subsidize the exploitation of male kin. Just as pineros previously experienced hardship while working for Anglo contractors, they continue to do so while working for Latinos.

Explaining the Latinization of Forest Work

The Role of Immigration Policy

Federal immigration policies have had a significant impact on the entry and settlement of Mexican immigrants in the Pacific Northwest. As detailed earlier, the Bracero Program played a critical role in bringing Mexican agricultural labour to Oregon during World War II. After the program ended, undocumented workers continued to pass through the Rogue Valley on their seasonal fruit-picking circuits, and some found work in tree planting. In 1986, the Immigration and Reform Control Act also greatly influenced immigrant settlement and occupational structure by granting amnesty to undocumented Latinos, many of whom had been doing agricultural work for a number of years. This legalization enabled some older immigrants to become forest contractors. Because only immigrants who had entered the United States before 1982 could apply for amnesty, however, a large number of workers still remain undocumented. This population, in addition to those immigrants who arrived after 1986, constitutes the majority of forest management workers in southern Oregon today.

Due to the legal, economic, and social tensions that surround immigration politics, it is not possible to prove that contractors *knowingly* hire undocumented workers. Plenty of evidence, however, suggests that employers simply look the other way when it comes to documentation status. Current employment policy dictates that contractors are only required to look at the I-9 (Employment Elegibility Verification) forms presented by workers and to refuse documents that appear to be "obviously fraudulent."[50] Yet forest contractors in southern Oregon rarely report false documents. As one contractor noted: "All I'm responsible to check is that they [workers] have a document that lets them work in the US. To me, they have to present an immigration and social security card. And then with that, they are hired. But I am not one to say if it is good or bad. I'm not an immigration officer [laughs]."[51] Later, this contractor admitted that 90 to 95 percent of all forest workers are probably undocumented.

Federal immigration policies have shaped the Latinization of forest work in yet another way. The recent militarization of the US-Mexico border has prompted immigrants to travel less frequently to Mexico and to bring their families to settle in the United States on a more permanent basis. As sociologists Douglas Massey and Wayne Cornelius have noted, recent US immigration policies have, ironically, led to more permanent settlements of Mexican immigrants in the United States. This trend is clearly evident in the Rogue Valley. As one forest worker put it, "Everyone seems to be coming with their wives and their children. Plus, immigration laws are getting really, really hard. So this coming in and out, it's difficult for families. So we've seen more workers bringing their families. And stay. Hopefully to stay. Everyone's wishing, for their wives' sakes, that they can stay."[52]

The Role of Federal Set-Aside Programs

While immigration policies set the broad contours of when Latinos could begin to enter into contracting, other programs helped facilitate their entry and early success. In particular, the Small Business Administration's 8(a) set-aside program, which reserves federal contracts for minority-owned businesses, played a pivotal role in levelling the playing field between new Latino entrants and older, well-established Anglo firms. Each fiscal year, the Washington office of the Forest Service sets procurement preference goals requiring federal agencies to negotiate socio-economic goals with the Small Business Administration. For a number of years, the Forest Service has been expected to award 10 percent of the total dollar value of all contracts to small, disadvantaged, and minority-owned businesses.

Since 1990, eleven Latinos in the Rogue Valley have been certified as 8(a) contractors. Most of these set-aside beneficiaries are long-established Latino contractors who have successfully used the program to build strong relationships and reputations with contracting officers in the Forest Service.[53] Firms that get certified under the 8(a) program can retain their status for a period of up to nine years. One contracting officer reflected on the success of established Latino contractors:

The Hispanic group has always felt very comfortable with our contracting officers. Our office has always been very helpful in getting them the proper information. I can't say for certain, but I'm pretty sure that southern Oregon has carried the region in terms of meeting 8(a) goals ... Our Hispanic contractors have developed so well out of the 8(a) program. They do not just plant trees. They have diversified out to weeding native grasses, gopher baiting, trapping, so much more. They do most of the nursery work too.[54]

The 8(a) program is less helpful to newer Latino contractors, as many are not yet certified under the program and are competing against older firms that still retain their 8(a) status. While the set-aside program may help these newer entrants as they gain certification in the future, its effect will likely be limited when compared to the boost given to the first wave of Latino contractors: firms getting 8(a) certification today must compete against previously certified participants, while those who entered the program during the early 1990s faced no such competition over set-aside contracts.

The Role of Kinship Networks
Kinship networks between labour contractors, foremen, and workers have also played a central role in the Latinization of forest work. While these networks were not available to the first Latinos to enter forest work in the early 1970s, they became central to the recruitment of subsequent waves of immigrant workers. By the early 1980s, Anglo contractors in the Rogue Valley had begun to appoint Latinos to supervisory positions as foremen, with the responsibility of hiring workers. As the case of Rubén Gómez reveals, Latino foremen actively recruited their friends and relatives onto forest labour crews, which rapidly became ethnically homogeneous. As these early foremen later became contractors, they continued the practice of hiring workers from their social networks. Labour recruitment through kinship ties also reduced contractors' costs and their need to engage formal recruitment processes.

While access to kinship networks makes it easier for contractors to recruit workers, it hinders the ability of pineros to seek improvements in wages and working conditions. Family ties also prevent pineros from lodging complaints about labour abuse and reinforce power asymmetries between worker and contractor, and between undocumented immigrant and legalized resident.[55] Other studies of immigrant workers reveal similar dynamics that connect labour productivity and compliance with the disciplining role of kinship obligations.[56]

The Role of Changes in Forest Management

Early Latino settlers noted that opportunities in tree planting influenced their decisions to settle in the Rogue Valley during the early 1970s. The prevalence of the word *reforestation* in the names of numerous contracting companies illustrates the centrality of tree planting to forest work through the 1980s and early 1990s. The demand for reforestation was closely tied to the amount of logging on public lands, and as more trees were felled, more were planted. In Oregon, the amount of timber harvested on federal lands increased consistently through the early 1980s and peaked at 4.926 million board feet in 1988 (see Figure 3.7). Peak levels of reforestation followed in 1991, with over 132,857 acres being replanted on Forest Service and Bureau of Land Management lands in Oregon (see Figure 3.8). Cutting on federal land from the 1970s through the 1980s thus fuelled a thriving reforestation sector.

A number of political, economic, and legal developments led to the demise of logging on federal lands in the western United States by the mid-1990s and halted the growth of the reforestation industry. Beginning in the early 1960s, growing concern about ecological health gradually culminated in the rise of a strong, domestic environmental movement.[57] Public outcry against clear-cutting on federal lands and an increasing resistance to utilitarian models of resource management eventually led to environment-oriented federal and state legislation.[58] Economic factors also contributed to the decline of logging in the West. In the late 1970s, operations began to relocate to the South as increasing competition, mechanization, and technological inefficiency rendered many of the large-saw mills of Washington, Oregon, and California defunct, led to unemployment among timber workers, and reduced the cut from public lands in the Pacific Northwest.[59] For example, between 1984 and 2004, employment in the logging industry in Oregon declined from approximately thirteen thousand to eight thousand employees.[60] With the decrease in timber harvesting from the early 1980s on, reforestation work also decreased by an estimated 10 to 25 percent.[61]

The final blow to cutting on federal lands came with Judge William Dwyer's halt on logging in spotted owl habitat in 1991. The subsequent passage of the Northwest Forest Plan by the Clinton administration in 1994 led to severe declines in logging on federal lands in western Washington, western Oregon, and northern California. Specifically, Alternative 9, the selected option for the Northwest Forest Plan, reserved millions of acres from logging. This included 7,320,600 acres of congressionally reserved areas; 7,230,800 acres in late-successional reserves; 1,477,100 acres of lands administratively withdrawn; and 2,627,500 acres

Figure 3.7

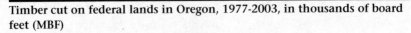

Timber cut on federal lands in Oregon, 1977-2003, in thousands of board feet (MBF)

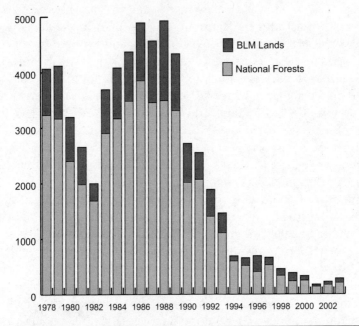

Source: Oregon Department of Forestry, "Oregon Timber Harvest, 1962-present." http://www.oregon.gov/ODF/STATE_FORESTS/FRP/Charts.shtml#Downloads.

in riparian reserves.[62] The dramatic decrease in logging on federal lands also led to the decline of manual reforestation activities; as fewer trees were cut down, fewer were required to be replanted. Indeed, Forest Service spending on reforestation contracts decreased by nearly 50 percent, from $44 million in 1992 to $24 million in 1999.[63] Reductions in timber extraction thus affected not only white, rural logging communities, but also reforestation workers.

With national forests no longer serving as prime suppliers of timber, many resource management policies have come to prioritize ecosystem health and fuels reduction. In August 2003, President George W. Bush stood on a hillside charred by the Quartz fire in southwestern Oregon. Flanked by members of the Secret Service and surrounded by a select few from the Forest Service and Bureau of Land Management, Bush announced his Healthy Forests Initiative, a policy forged largely in response to the sumer of 2002, one of the most catastrophic fire seasons

Figure 3.8

Acres planted on Forest Service and Bureau of Land Management land in Oregon, 1969-98

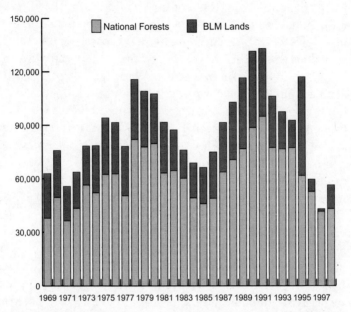

Note: Acres planted for 1977-2003 include acres seeded and acres of windbarrier planting. Acres planted for 1969-76 include acres seeded but exclude acres of windbarrier planting.

Source: US Department of Agriculture, Forest Service and Southern Regional Extension Forestry, "Tree Planting in the U.S. – Reforestation, Nurseries and Genetic Resources," 2010, http://www. rngr.net/resources/tpus.

on record.[64] The Healthy Forests Initiative – or, in its more current form, the controversial Healthy Forests Restoration Act – aims to promote "healthy" forests by reducing fire danger through fuels reduction programs and by restoring lands to an ostensibly more natural state.[65] The policy also claims to promote "healthy" communities by protecting people's homes and property in wildland-urban interfaces and providing employment opportunities, primarily through fuels reduction and ecosystem management projects, in the woods. In addition to the Healthy Forests Restoration Act, the National Fire Plan, launched in 2001, also seeks to increase fire suppression capabilities, reduce fire hazards, restore fire-adapted ecosystems, and create economic benefits for rural communities and businesses.[66]

With the rise of the Healthy Forests Restoration Act and the National Fire Plan, the bulk of work on public lands in the western United States shifted from logging and reforestation to fuels management, stand improvement, habitat enhancement, and wildfire suppression, often all lumped together under the category of "ecosystem management work" and under the umbrella of the "forestry services industry." Ecosystem management has been defined as an approach to land management that enhances, protects, and maintains natural ecosystem functions and biological diversity.[67] Specific ecosystem management activities include heavy equipment work such as stream clearing and road decommissioning, manually intensive work such as hand-thinning dense underbrush, piling and burning brush, and planting trees, as well as more survey-oriented work.[68] Between 1983 and 2003, this industry grew by 107 percent in Oregon.[69] During the 2001 fiscal year, the Forest Service procured over $10.3 million of goods and services from National Fire Plan funds, and the Bureau of Land Management procured over $14.8 million in Oregon and Washington alone. Of this combined $25.1 million, $8.2 million was used for thinning work.[70]

Latino forest labour contractors and their immigrant crews were quick to move into much of the manually intensive, ecosystem-oriented work on federal lands. More recently, Latino forest labour contractors have also diversified into running fire crews. Contractors must apply for a separate licence to run fire crews, and workers have to complete required training. Fire fighting on federal lands can be highly rewarding (and dangerous) work since the government pays for transportation, food, lodging, and overtime. Fire fighting is also highly seasonal and part of a larger income-earning strategy for year-round work in the natural resources sector.

There are three main reasons for the shift of Latinos from reforestation into ecosystem management work. First, given their experience running reforestation crews, Latino forest labour contractors were already "primed" to take on a new regime of manually intensive land management activities. Second, in southern Oregon, the Forest Service and Bureau of Land Management reserved a large portion of service contracts, including contracts for restoration work, in the form of 8(a) contracts. Finally, with the exception of Latino forest labour contractors and their crews, there was no viable workforce in place to take on the range of newly introduced ecosystem-oriented activities. As such, even as the number of reforestation contracts on federal lands plummeted after the implementation of the Northwest Forest Plan, the number of Latino forest labour contractors and the availability of ecosystem-oriented work grew during the 1990s. This trend suggests that Latino contractors, both in the Rogue Valley and in Oregon as a whole, have been quite successful in transitioning from reforestation/tree-planting contracts to the manually intensive ecosystem management contracts now put out by the Forest Service and Bureau of Land Management.

Overall, federal policies and immigrant social networks have contributed to the Latinization of manually intensive forest management, and they have done so in contingently specific ways. For example, the Small Business Administration's set-aside program aided the entry of Latinos into contractor positions, but it could not do so until the passage of the 1986 Immigration Reform and Control Act. Similarly, immigrant social networks did not play a significant role in the initial entry of immigrants into southern Oregon or into forest work. However, they did account for the entry of later immigrants, with recruitment from networks of friends and relatives becoming more frequent as pineros moved up the ranks of the reforestation industry into foreman and contractor positions. Finally, the settlement of Latino families in the Rogue Valley during the 1990s depended in part on increasing opportunities in forest management work, but also on federal policies that reduced circular migration between the United States and Mexico.

The Latinization of forest work has been produced through a confluence of factors, ranging from federal land-management policies and small-business programs to the seasonality of rural economies and the strength of immigrant networks. Any historical analysis of forestry in the United States thus needs to extend beyond the decline of logging on federal lands and the focus on the spotted owl controversy, or the attention given to elements of the "natural world" like trees, rivers, and endangered species. While political ecologists and environmental historians have made invaluable contributions to our understandings of the social production of nature, critical gaps remain to be filled. The labour exploitation of pineros is due in significant part to the fragmented nature of policy making: decisions and debates about the "natural" environment take place separately from discussions of immigration policy and the implementation and enforcement of labour laws. Only by taking into account the inextricable relationships between labour and landscape and by accepting Latino immigrants as integral players in forests and rural communities can we hope to have land management policies that promote both forest health and environmental justice.

4
The Marginality of Forest Workers

Juan Gonzales and Raul Padilla crossed the US-Mexico border in 1990. They were looking for work to support their extended families in Michoacán, Mexico. Juan was thirteen and Raul was fifteen. Both had dropped out of school in sixth grade in order to make the journey to El Norte. Like many other young, undocumented workers, Juan and Raul joined the migratory fruit-picking circuit and travelled between the agricultural fields of Fresno, California, and the orchards of the Yakima Valley in Washington. On one of these trips, Raul learned through a friend about opportunities to work in the woods, planting trees around Medford, Oregon. He remembers thinking: "Tree planting was going to be no big deal. We were just going to the mountains to plant five or six pines."[1]

The reality could not have been more different. Raul was packed into a van with sixteen other people, driven to a remote and steep site, and ordered by his foreman to haul bags containing 250 saplings and weighing fifty pounds each. Raul and other planters used a hoedad, an extended shovel-like tool, to dig holes in the rocky terrain and plant evergreen saplings. The crew started their drive to the planting site at 3:00 a.m. and worked until 4:00 p.m. with only a short break for lunch. By the time they had cleaned their equipment and returned to town, it was well past dusk. Depending on how much they planted and their level of seniority, workers were paid $12 to $15 an hour, with no benefits. Their wages did not include the six hours spent driving to and from the work site. If workers were injured on the job, they were simply let go. Raul spoke of being unable to work for two months due to a back injury incurred while planting. He received no compensation even though all workers, whether documented or not, are entitled to workers' compensation. Raul did not report his injury for fear of being unable to find future work. He says he is lucky to have recovered so he could plant again.

Raul's situation parallels the experience of numerous non-white immigrants in various arenas of the US private sector, including farm labourers and service workers. Ultimately, the labour marginality of pineros, in contrast to other groups of forest workers, stems not only from the isolated nature of forest work itself – work performed in remote areas and therefore often invisible to labour inspectors and the general public alike – but also from disadvantages related to the immigrants' race, lack of citizenship, family obligations, and limited English proficiency. These factors often limit pineros' ability to improve their working conditions or move out of undesirable jobs.

Sociologists and other scholars have discussed disadvantaged workers primarily in terms of "bad jobs," characterized by low wages, dangerous working conditions, no benefits, and part-time or contingent labour arrangements.[2] In contingent work, full-time jobs with benefits are replaced by contracted and subcontracted work, enabling firms to maintain the same level of productivity without having to provide their workers with benefits or job security. The growth of precarious employment in the United States and other industrialized countries is related to a number of factors, including increased global competition, just-in-time production, lower profit margins, and rapid changes in consumer demand and production technologies. Both low-income and immigrant workers and white-collar professionals in the United States are now often employed as contingent labour, including part-time, seasonal, and subcontracted work.[3] Overall, employer accountability is increasingly tenuous as workers across the spectrum find themselves in highly insecure forms of employment, accompanied by the constant fear of losing their jobs.

Due to the seasonality of work in natural resource industries, contingent work prevails in logging, reforestation, and ecosystem management.[4] In 2003, although forest workers in Oregon earned a median wage of $11.73 per hour, over half earned less than $4,355 annually, and more than 85 percent earned less than the federal poverty level for a family of four.[5] In recent years, severe cuts to Forest Service budgets have accelerated the outsourcing of operations such as nursery work in southern Oregon to immigrant labour contractors. Most contract workers on federal land have seasonal jobs and no benefits. Moreover, contracting arrangements conveniently transfer the risks and responsibilities involved with hiring labour and doing forest work from federal land-management agencies to contractors. Such arrangements often enable contractors to lower their bids by driving workers' wages down.

Contracting and subcontracting arrangements also weaken the power of workers. Contracted workers have two bosses but do not receive the same benefits, opportunities for advancement, and job security to which "real employees" of their de facto employers are entitled.[6] In fact, sub-contracting hinders workers' ability to organize and demand improved labour conditions because it disperses workers among many different employers and limits opportunities for face-to-face communication between workers within any particular industry.[7] In forestry, land management agencies consistently deny responsibility for workers who are treated poorly or left unpaid while working on federal lands. The Forest Service and Bureau of Land Management claim that it is the contractor's responsibility to adhere to labour and employment laws, and that enforcement of such laws is the onus of the Department of Labor and not that of land management agencies.[8]

This claim is problematic for two reasons. First, it raises the question of how Department of Labor officials are supposed to access remote and forested work sites when, often, only inspectors from the Forest Service and Bureau of Land Management know where contracts are being carried out and are on-site to witness workplace violations. Second, the vulnerable legal status of many pineros serves to dampen the likelihood that they will complain about labour violations. Relations of co-ethnic exploitation and the dire political context for undocumented immigrants in the United States today fuel a climate of fear and silence, exacerbate labour exploitation, and drive immigrant workers further into the shadows.

Comparing Loggers, Former Tree Planters, and Pineros
Despite the predominance of Latino forest workers since the mid-1980s, the media, environmentalists, land managers, and natural resource policy makers have paid scant attention to this workforce. The poor working conditions and continued political invisibility of pineros contrast sharply with the situations of Anglo forest workers in the Pacific Northwest, including the contract and co-operative reforestation workers in the 1970s and early 1980s, and loggers since World War II. Unlike their immigrant counterparts today, tree planters in the mid-1970s and early 1980s had legal recourse in cases of contractor abuse and failure to pay. Members of tree-planting co-operatives also played an active role in their organizations and directly communicated with government agencies such as the Forest Service to demand dignity and respect on the job.

Between World War II and the early 1980s, many loggers in the Pacific Northwest enjoyed the benefits of union representation, full-time work, and company benefits. If loggers were small, independent operators (also known as "gyppos"), they generally had access to sufficient amounts of well-paid work.[9] Workers injured on the job, moreover, were guaranteed workers' compensation and, if a company employee, disability benefits. Out-of-work loggers were able to claim unemployment insurance.[10] Although the employment security of loggers has changed dramatically since the mid-1980s, federal and state programs have recognized this displaced workforce and attempted to provide support through various initiatives, including the Jobs in the Woods program of the 1994 Northwest Forest Plan.[11] The media has also, on occasion, prominently covered the vicissitudes of the timber industry and of pitched battles between loggers and environmental activists, thus keeping the concerns of these forest workers, and their industry, in the public eye.[12] The same cannot be said of the visibility of the reforestation sector or its predominantly Latino immigrant workers.

Loggers

Logging has a rich and complicated history of labour organizing and worker representation. From the more radical Wobbly organizing of pre-World War I transients, to the Loyal Legion of Loggers and Lumbermen during World War I, to tensions between the American Federation of Labor and the Congress of Industrial Organizations and the founding of the International Woodworkers of America in 1939, loggers have been part of a larger struggle for organized labour in the United States.

Disgruntled loggers traditionally responded to dangerous working conditions, coercive management, and crude accommodations by simply quitting and moving on: the timber industry workforce was highly transient. Changes in logging technology during the latter part of the nineteenth century, however, necessitated a more disciplined and efficient workforce. Historian Richard Rajala aptly notes that operators could not "afford to let expensive machinery lie idle or be operated at less than maximum capacity while loggers drifted in and out of camps."[13] Coastal logging entrepreneurs thus had a vested interest in improving camp conditions and bringing some stability to their workforce. The rise of welfare capitalism among timber operators during the 1920s inaugurated numerous efforts to thwart radical tendencies and create a docile, compliant, and loyal labour force. Measures included introducing ideologically conservative literature into logging camps, providing group insurance for employees, hosting company picnics, and constructing company towns for loggers' families.[14]

Labour conditions for most loggers improved considerably in the post-World War I period as electricity, decent bunkhouses, and spring mattresses were introduced to logging camps. However, not all who worked in the timber industry benefited from these improvements or shared equal workplace risk. Although logging was predominantly a white occupation, it was structured through racialized hierarchies such that non-white workers were at a disadvantage. African Americans working in northern California logging camps during the early 1920s held the most dangerous positions and lived in segregated and inferior camps.[15]

Overall, the success of welfare capitalism to sow stability (and loyalty) among timber workers was limited. Depression-era unemployment took a harsh toll on loggers and sawmill workers, and between 1929 and 1933, employment in Oregon's logging industry fell from 32,532 to 13,007 workers, a decline of 60 percent. During the same period, total annual wages in the state's lumber industry dropped from $51 million to $11 million, a decline of 77 percent.[16] Unemployment and the fall in wages, accompanied by deteriorating working conditions, spurred walkouts and renewed agitation among loggers. Seething with discontent, an estimated thirty thousand loggers and sawmill workers in the Douglas fir region struck in May of 1935.[17] This large-scale collective action temporarily crippled the timber industry and strengthened union membership in the years leading to and following World War II.

The post-World War II boom in timber demand brought some stability to the lumber industry as federal lands were opened up for harvesting. Between 1950 and 1970, timber workers in the Pacific Northwest prospered, and company loggers and gyppo operators sustained a heavily timber-dependent economy in the Pacific Northwest. During the 1960s, for example, nearly 15 percent of Washington's employment was in timber-related jobs, and until 1979, the industry employed 160,000 workers.[18] The opening up of federal lands for timber harvests also led rural counties in the Northwest to become heavily dependent on timber sale receipts to finance schools and roads.

Ironically, World War II also set the stage for the pacification and gradual decline of US labour. During the war, both the American Federation of Labor and the Congress of Industrial Organizations adopted no-strike policies, and most workers supported the war. Anti-strike policies ultimately ensured that wartime production would continue without disruption. After the war, the 1947 Taft-Hartley Act gutted union power and led to the reign of business unionism. The act gave employers "right to work" policies, banned secondary boycotts and sympathy strikes, and made the union certification process much less flexible and less responsive to labour.[19]

From the 1970s on, consolidation and mergers within the timber industry, labour-saving technology, and a shift in operations to the southeastern United States led to declines in the viability of smaller gyppo operators.[20] By the mid-1980s, union density waned dramatically as large timber companies such as Weyerhaeuser cut wages, laid off workers, and shifted to non-union, independent contractors. The decline in the number of unionized loggers has also meant that most loggers today do not receive benefits due to their status as independent contractors. With the demise of logging on federal lands, employment for timber workers in the Pacific Northwest has become even more precarious.[21]

In light of significant blows to their livelihood, Pacific Northwest loggers have been quite vocal about forest management decisions.[22] Media coverage on the spotted owl during the mid-1990s framed the issue in terms of a conflict between environmentalists and loggers, and both state and federal politicians were quick to acknowledge (if not alleviate) the plight of displaced timber workers.[23] One of the five key principles of the Northwest Forest Plan was to "never forget human and economic dimensions of issues," and the plan included an initiative for providing economic assistance and job retraining to displaced timber workers, communities, and others who were adversely affected by reductions in the size of the timber program. Such shows of state support reveal that even though logging may be a disappearing occupation in the Pacific Northwest, timber workers are still recognized as an important constituency.

For multiple reasons – including the historical organizing of timber workers, the economic clout of the timber industry, cultural legacies rooted in logging, and the composition of loggers as a primarily white and citizen-based workforce – logging and loggers have continued to garner the attention of policy makers. Even as timber-related occupations are on the wane, loggers remain prominent in considerations about forest policy, debates over forest management, and media coverage.

Economic Marginality among Reforestation Workers

Most forest workers endure dangerous working conditions and receive low wages. Although contingent-labour arrangements contribute to the economic marginality of all forest workers, pineros are also marginalized in ways distinct from those of their white, US-born counterparts.

Safety

Safety on the job is a perennial concern for forest workers in logging and forestry services in Oregon. Nationwide, forest work accounts for 2 percent of all occupational fatalities, with 92 deaths per 100,000 workers.[24]

Table 4.1

Indicators of economic marginality among forest workers in Oregon

Industry and workers' ethnicity[6]	Safety (2004)*		Wages (2004, average)*		Benefits[4]		Structure of employment relations[5]	
	Fatalities[1]	Disabling claims[2]	Annual (Weekly)[3]	Issues	On paper	Issues	Standard	Non-standard
Logging[6] Primarily Anglos historically and currently	8 Rate: 103.7 per 100,000 workers	483 Rate: 6.3 per 100 workers	$37,504 ($721)	None	Company loggers: health care, disability, workers' comp., U.I.	Gyppo loggers have no benefits	Company loggers	Gyppo loggers and their employees are contract workers
Forest services/ support activities for forestry[7] Primarily Anglos 1970-1985 (tree planting)	0 Rate: 0	131 Rate: 3.3 per 100 workers	$22,832 ($439)	Ability to report labour violations Co-ops have no minimum wage	Workers' comp., U.I.	U.I. only available to citizens and legal residents		Contract workers or co-op members

Primarily Latinos 1985-early 1990s[8] (tree planting)	Wages not paid on time; Lack of payment for work done; Inability to report labour violations	Workers' comp.	Legal status and kinship ties keep most pineros from reporting injuries or claiming workers' comp.	Workers hired through labour contractors

* Safety and wage statistics reflect contemporary conditions for forest workers in Oregon. I was unable to find reliable statistics on fatalities, non-fatal injuries, and wages for tree planters during the 1970s.

1 Statistics on fatalities are based on industry codes for "Logging" (North American Industrial Classification System, NAICS 11331, previously Standard Industrial Classification SIC 2411) and "Support activities for forestry/forest services" (NAICS 11531, previously SIC 0851) (Department of Consumer and Business Services, 2005).

2 Data on non-fatal occupational injuries are based on cases by NAICS industry codes "Logging" (NAICS 1133) and "Support activities for forestry" (NAICS 1153) (US Bureau of Labor Statistics, 2004, "Numbers of nonfatal occupational injuries and illnesses by industry and case type, Oregon, 2004).

3 Annual wages per employee and average weekly wages are based on six-digit NAICS industry codes: "Logging-11331" and "Support activities for forestry-11531" (US Bureau of Labor Statistics, 2004, "Private industry by state and six-digit NAICS industry: Establishments, employment, and wages, 2004 annual averages").

4 Benefits include health care and disability for employees. I have also included workers' compensation and unemployment insurance in this section.

5 Standard employment relations refer to full-time, year-round work with an employer, usually with access to benefits such as health and disability insurance. Non-standard employment relations (sometimes called "contingent work") refer to part-time, seasonal, and contracted or subcontracted work, usually without access to any benefits (Kalleberg, Reskin, and Hudson, 2000).

6 "Logging" includes workers involved in "establishments and activities primarily engaged in the cutting and removal of timber from public and private lands" (Oregon Department of Consumer and Business Services, 2005: 1). For the purposes of this chapter, loggers (in the past and today) are primarily white, Anglo, male, and US citizens.

7 The Bureau of Labor Statistics classifies tree planting and ecosystem management activities under one industry: "Support Activities for Forestry/Forest services," which includes "activities such as forest firefighting and fire prevention, timber cruising, reforestation, and vegetation management" (Oregon Department of Consumer and Business Services, 2005: 1). For profiles of industries using NAICS/SIC codes in Oregon, see http://www.qualityinfo.org/olmisj/CEW.

8 From the mid-1990s on, work shifted to ecosystem management activities. Ecosystem management work includes heavy equipment work such as stream clearing and road decommissioning, manually intensive work such as hand thinning dense underbrush, piling and burning brush, and planting trees, and more survey-oriented work (Baker, 2003). Latino forest workers primarily do manually intensive ecosystem management work (see Chapter 4).

This injury rate far exceeds the rate for mining and manufacturing workers.[25] While all forest work is dangerous, logging has the highest fatality rate. In 2004, there were 103.7 deaths per 100,000 logging workers in Oregon, with a total of 8 fatalities, all among white workers.[26] The logging industry only has 0.05 percent of the employment in Oregon, but it accounts for 2.2 percent of the accepted disability claims in the state.[27]

Although there were no fatalities in the "forestry services" industry (which comprises forest and conservation workers, including tree planters) in Oregon during 2004, pineros' accounts and recent media reports on Latino floral-greens harvesters suggest that non-logging forest workers also suffer workplace fatalities. In a nine-month investigative series on Latino forest workers in the United States, reporters Tom Knudson and Hector Amezcua revealed that at least twenty-three pineros have died in van-related accidents since 2003.[28] In the Rogue Valley, many workers mentioned travelling to work in unsafe vehicles, typically with fifteen or more people squeezed into vans without seats or seatbelts. Transportation-related fatalities are also high among agricultural workers. In August 1999, thirteen farm workers were killed when their van collided with a semi-trailer in Fresno County, California. One month later, another thirteen workers were injured or killed south of Fresno when their van collided with another car. At the time of these incidents, 45 percent of work-related fatalities in agriculture were related to vehicles.[29]

The danger of forest work is also apparent in the prevalence of non-fatal injuries suffered by workers. National data from the US Bureau of Labor Statistics reveals that in 2004, loggers sustained sixteen times as many injuries as forest and conservation workers.[30] Among loggers, equipment operators sustained nearly three times as many non-fatal injuries as timber fallers, but timber fallers suffered more on-the-job fatalities. The most commonly reported causes of injuries among loggers were contact with outside objects (downed logs or branches, or equipment such as chainsaws) and falls. Finally, of the 810 reported injuries among loggers nationally, more than half involved white workers.

By contrast, in 2004, the Bureau of Labor Statistics recorded only fifty non-fatal injuries, nationally, for the Standard Occupational Classification category of "forest and conservation workers," of which forty were Hispanic. Data from other sources indicate that this is far too low an estimate.[31] Such discrepancies highlight the problem of using Standard Occupational Classification codes to get information on forestry services workers in particular. Industry-specific data is more reliable. For the same year in Oregon, 483 non-fatal disabling claims were accepted for the logging industry, with 6.3 of every 100 workers filing claims. Contact with outside objects accounted for the majority (29.2 percent) of disabling claims in logging, followed by falls (21.1 percent), bodily reaction (12 percent), and overexertion (8.7 percent).[32]

By contrast, 131 disabling claims were accepted for the forestry services industry in Oregon in 2004, with 3.3 of every 100 workers filing claims. The most frequent accident reported that year was falls, with a total of 25 claims. Twenty-four workers also claimed injuries due to being struck by objects, and twenty-three reported incidents of bodily reaction.[33] There was no corresponding data available on the race or ethnicity of workers who died or were injured in the logging or forest services industries in Oregon.

Tree planters in particular sustained non-fatal injuries during the early 1980s. The constant stooping and hacking at rocky soil to plant saplings often resulted in repetitive stress injuries such as carpal tunnel syndrome, acute and chronic lower back strain, and slipped discs.[34] Pineros in the Rogue Valley frequently talked about work-related injuries, including being cut by chainsaws and struck by tree limbs. Yet while many admitted to being injured, most did not report their injuries for fear of losing jobs or being called "crybabies" by other workers. This "grin and bear it" attitude towards injuries belies the reality that most immigrant labourers are not provided with safety equipment such as goggles or chaps by their foremen or contractors. By not reporting unsafe working conditions or injuries, many pineros effectively absolve contractors and foremen, who are also often family members, from responsibility for workplace injuries.

Mauricio Benavides recounts being injured while working for his father's company:

It was in the morning and we were down a canyon and up another one. And it was my fault. I was slacking and the chainsaw slipped and sliced my leg. So then ... it basically just went "rrrrippp." And I kneeled over and blood started coming out. I took my shirt off and tied it. The guys didn't say, "We'll carry you up" ... And the foreman said, "If you are heading up, take those gallons of gas with you, will you?" And I did. I dragged my leg up. It took three hours to make it up. I got to the top and fainted ... [Americans or Anglo workers] would have probably got workers' comp. And a helicopter. But we [Latinos] are poor people. We train ourselves not to be crybabies.

Mauricio displays the keen awareness among pineros that they are doing hazardous work with less protection from accidents than Anglo forest workers. At the same time, Mauricio's recognition of this disparity does not spur him to demand restitution and justice. Instead, the disparity is rationalized away by adopting a certain bravado in working through pain and seeing workers' comp as a luxury rather than a right. Moreover, if Mauricio *had* filed for workers' compensation, he would have exposed his father's business to official government scrutiny and jeopardized the employment of other undocumented relatives on the crew.

Wages
Wage and employment data from the US Bureau of Labor and Statistics indicates that the *national* median and mean hourly wages for forest workers, according to specific Standard Occupational Classification categories, are as follows: "logging equipment operators," $13.91 and $14.28, "fallers," $13.64 and $15.26, and "forest and conservation workers," $9.46 and $11.19. Wages for logging work (on private land) are particularly low because the national mean and median hourly wages include logging workers in the southeastern United States, who earn significantly less than their counterparts in the Pacific Northwest. For example, in Alabama, "logging equipment operators" earn a median hourly wage of $12.68 and "fallers" earn $11.48. By contrast, in Oregon, "logging equipment operators" earn $16.65.[35]

State wage estimates for private industry also reveal higher wages in logging than in forestry services. In Oregon, the annual average wage per employee in the logging industry stood at $37,504, and the average weekly wage was $721. By contrast, employees in the forestry services industry earned an annual average of $22,832 with an average weekly wage of $439.[36] Overall, although logging workers earn more than those in forestry services, neither group earns very much.

Wages for work rendered to federal agencies such as the Forest Service or Bureau of Land Management are determined by the Service Contract Act, and vary by state. In Oregon, in 2004, wages prescribed by the act for work on federal contracts stood at $12.90/hour for thinning, $11.69/ hour for tree planting, $8.30/hour for slash piling, and a "bonus" of $5.29/hour in lieu of benefits.[37] In addition, overtime wages are required on federal contracts as stipulated by the Contract Work Hours and Safety Standards Act. Pineros are also covered under the 1983 Migrant and Seasonal Agricultural Worker Protection Act, which is the major federal law that deals with agricultural employment. The law was enacted "to protect migrant and seasonal farm workers on matters of pay and working conditions, to require farm labour contractors to register with the US Department of Labor, and to assure necessary protections for farm workers, agricultural associations, and agricultural employers."[38]

Together, these laws require that "employers provide workers with information about legal entitlements, detailed accounting of their pay, transportation in safe, insured vehicles, and other provisions."[39] Yet employers frequently violate labour regulations. While forest workers earn between $10 and $12 an hour, crew foremen often (illegally) deduct transportation charges from their pay and thereby bring wages below the stipulated minimum. Such practices are violations of the Contract Work Hours and Safety Act and Migrant and Agricultural Seasonal Worker Protection Act.

Although Anglo tree planters in the 1970s and early 1980s also experienced situations in which labour contractors did not pay, some of these workers took action to reclaim unpaid wages. In Hal Harztell's book about the Hoedads, tree planter Lowell Rose recounts how he and other workers had been "ripped off" by a contractor. The tree planters first went to the contractor's home to demand their money and then turned to the government for help:

> We went to talk to the contractor. His wife was there and she told us that he had fought in Vietnam and that he would never rip us off. Then she said they didn't have the money anyway, and if they had to give us our bonus they would go bankrupt. Anyway, the contractor finally came and threw us out of his house. We wondered what we were going to do next. We filed a complaint with the Bureau of Labor. The foreman backed us up. The contractor had fired him. It took a couple of months before our case went to Salem for a hearing. Basically, they pinned him [the contractor] against the wall ... He finally copped out and said he would give us half, and that if we wanted more we would have to take him to court. None of us had the money to take him to court, so we settled on half, and that was the end of that contractor. As far as I know, he went out of business.[40]

Lowell's experience shows that Anglo planters were aware of their rights, had the skills (particularly the ability to read, write, and communicate in English) and knowledge to file complaints, and had access to state labour officials. The willingness and ability of these forest workers – and their foreman – to confront the labour violations of a contractor stands in stark contrast to the silence of most pineros today. Anglo re-forestation workers, unlike pineros, were also less bound to contractors (or foremen) through kinship ties, and they did not fear suffering repercussions as dire as deportation. Indeed, Lowell attributes the inability of Anglo planters to reclaim *all* their wages only to a lack of funds needed to initiate court cases.

It is important to note that wage exploitation was voluntary for co-op members whereas for the pineros who followed, it was not. Tree planters who belonged to reforestation co-operatives did not initially receive a minimum wage for their work. As co-operative members, they determined their own wages. Scholar and former planter Gerry Mackie notes that co-operative members actually earned quite a bit more than minimum wage during the late 1970s: "Wage returns increased every year, and by 1980, top co-operative forest workers were earning $25 an hour and $25,000 a year and more (in 1980 dollars)."[41] But high wages such as those noted by Mackie were the exception, not the norm. Still, even in co-operatives in which members earned lower wages, the coercion within the employer-employee relationship was absent. Co-operatives often decided to "self-exploit" and offer lower bids on federal contracts in order to compete against contractor-run firms.

Overall, loggers, tree planters, and pineros experience varying degrees of economic marginality. In terms of safety, loggers sustain the most labour fatalities and non-fatal injuries. But loggers also get paid higher wages and can claim workers' compensation and unemployment insurance. Full-time company loggers also have access to disability benefits and health care. However, gyppo loggers, like other contingent workers, do not have benefits. Contingent-labour arrangements in logging are becoming the norm within the industry and increase the economic marginality of loggers.[42]

Whether performed by Anglos during the 1970s or by pineros today, tree planting and ecosystem management have historically been structured through contingent-labour arrangements. Anglo tree planters, however, had greater recourse to address labour violations than do their Latino counterparts today. Despite the marginal nature of reforestation work, Anglo workers were able to secure wages for their work, claim workers' compensation for injuries, and file for unemployment insurance. The same cannot be said for the pineros who do similar types of forest work today.

Pineros have no benefits and face numerous problems related to compensation, including unjust compensation (being paid for only some of the hours worked), untimely pay, and in extreme cases, not getting paid at all. Due to their often precarious legal status and complicated social and kinship ties to employers, most pineros do not claim workers' compensation or unemployment insurance. Other research has found significant differences in the job quality of forest workers based on worker ethnicity, with Hispanics doing a disproportionate share of manually intensive forest work and being more likely than whites to work seasonally, away from home, and for companies that do not offer health insurance.[43] The preponderance of evidence thus indicates that pineros are the most economically marginal of the three categories of forest workers.

Social Invisibility/Marginality among Forest Workers

Media Coverage

Media coverage influences the social visibility and public awareness of particular groups and issues.[44] The media can play a significant role in shaping the public's understanding of forest politics and keeping them on the agenda of policy makers. For example, in 2002, newspapers across the country extensively covered the Biscuit fire and prominently highlighted the problem of catastrophic fires in the western United States. Partially in response to this coverage, George W. Bush launched the Healthy Forests Restoration Act in 2003, seeking to mitigate catastrophic fires on federal lands.[45]

How, then, have different groups of forest workers been covered in the media? Do some groups get more public exposure via the media than others? Examining coverage of loggers, former tree planters, and pineros in the *New York Times* and in southern Oregon's largest circulating daily, the *Medford Mail Tribune*, helps answer these questions.[46] In the past two decades, "loggers" and "logging" received by far the most media coverage in both national and local newspapers (see Table 4.2). The *New York Times* (*NYT*) published thirty-eight relevant articles on loggers, while the *Medford Mail Tribune* (*MMT*) mentioned loggers and logging in at least sixty-one articles. Articles in the *NYT* included stories on the timber wars during the mid-1990s, loggers' opinions on environmental protection for fish and wildlife, the political involvement of loggers in the Northwest, the displacement of loggers as a result of the Northwest Forest Plan, government aid and retraining for displaced timber workers, debates on clear-cutting, and economic vicissitudes within the timber industry.

The *MMT* focused primarily on the passionate debates between environmental activists and timber advocates over logging on federal lands

Table 4.2

Media coverage of loggers, tree planters, and pineros

Media coverage (number of articles)

Forest worker	Type of forest work	New York Times (1975-2005)	Medford Mail Tribune (1997-2005)
Loggers (white, male, US citizen, Anglo)	Logging	38	61
Tree planters (white, primarily male, some female, US citizen, Anglo)	Tree planting	4	0
Pineros (non-white, male, immigrants, legal residents and undocumented, Latino)	Ecosystem restoration, ecosystem management	0	• 0 about pineros • 18 about ecosystem restoration, but only 2 mention workers (both Anglo workers)

in southern Oregon. After the wildfires of 2002, many of the *MMT* articles also specifically covered the issue of salvage logging fire-killed trees. Indeed, the two most prominent actors in media coverage on forest management in the Rogue Valley were loggers and environmental activists. Of the 126 articles about logging, forest thinning, and ecosystem restoration, 61 articles mentioned loggers and 46 covered environmentalists. The most prevalent images of "forest actors," moreover, were those of white environmental activists, whether they were rallying at public protests, challenging timber advocates, or blocking logging trucks from getting to work. Overall, both the *NYT* and *MMT* presented loggers and environmentalists as actively engaged and central players in debates over forest management in the Pacific Northwest and in the Rogue Valley.

By contrast, the only newspaper coverage on the reforestation industry consisted of four pieces in the *NYT* between 1984 and 1987. Two of these were letters to the editor regarding the proposed use of Youth Conservation Corps workers to plant trees on public lands. The other two pieces only briefly referenced reforestation in the western United States. Interestingly, one of the letters to the editor was from Gerry Mackie, who at the time was a member of the Hoedads co-operative.

Mackie argues that youth workers would displace other tree planters, but he fails to mention any details about reforestation workers. The publication of Mackie's letter shows that some members of reforestation co-operatives did publicly voice their concerns about forest work.

Any mention of reforestation workers or the reforestation industry in the *NYT* disappears after 1987, coinciding with the rise of a non-white and immigrant labour force.[47] The *MMT*'s lack of coverage on tree planting between 1997 and 2005 is not particularly surprising because planting declined along with the reduction of timber harvests on federal land during the early 1990s.[48] Given limited access to *MMT* archives, I was unable to find coverage on tree planting from earlier years when it was a more prominent type of forest work. The only relevant pieces about immigrant forest workers concerned Southeast Asian mushroom harvesters in Oregon; one of these articles was published in the *NYT* and two others appeared in the *MMT*. However, mushroom harvesters do not do the same type of work as pineros contracted on federal lands. Finally, only two of the *MMT*'s eighteen articles about ecosystem management mentioned forest workers, both dealing exclusively with Anglos.

One notable exception to the lack of news coverage of pineros is a November 2005 series by Tom Knudson and Hector Amezcua in the *Sacramento Bee*. The series largely focused on H-2B guest workers, who are more prevalent in the southeastern United States, and led to congressional hearings on their labour conditions. The spotlight on guest workers, however, resulted in less attention being paid to undocumented pineros on federal lands in the Pacific Northwest.[49] But the *Sacramento Bee* coverage was important in highlighting the plight of immigrant workers in forestry, and it helped to catalyze coverage in *The Oregonian*, the *Seattle Times*, and the *New York Times*. The *Oregonian* and *New York Times* pieces touched on the prevalence of immigrant (primarily Latino) forest fire fighters, while the *Seattle Times* profiled immigrant brush harvesters on the Olympic Peninsula of Washington.[50] These articles indicate that regional newspapers are finally taking note of the demographic changes in the forest labour force, although the coverage is often limited to one story or one series, with no sustained coverage over time.

With the exception of the *Sacramento Bee* series, media coverage about pineros has generally been sparse. The complete absence of coverage about pineros in the *Medford Mail Tribune* is especially ironic given that Latinos make up the vast majority of the forest management workforce and contractor population in the Rogue Valley. The paucity of coverage on pineros in forest work not only reveals their social marginality in relation to loggers and Anglo tree planters, but it also contributes to public ignorance about the Latino forest labour force and the work in which they are engaged.

Organizations and Policy Engagement

Loggers

Even though loggers in the Pacific Northwest have faced severe job displacement, and even though the influence of unionized, company loggers has waned considerably, loggers continue to be a highly visible group in debates over forest management. In the Rogue Valley, independent loggers and timber advocates, including members of the Southern Oregon Timber Industries Association (SOTIA) and the Association of Oregon Loggers (AOL), turned out at rallies in Medford, Oregon, to publicly support salvage logging operations in the Rogue River-Siskiyou National Forest. The numerous lawsuits and protests over salvage logging the Biscuit burn only further spotlight loggers and the timber industry on one side and environmentalists on the other.

In many instances, the high profile of loggers may be attributed to the roles and resources of timber companies and professional organizations rather than the clout of smaller operators. Long-time gyppo logger Rob Pointer noted how the AOL was only out to help big loggers:

> See, we belong to the Association [of] Oregon Loggers. We joined it for one reason only. You could get workers' comp for a little bit of a discount ... Other than that there was no advantages. For the small logger, there was no advantage at all. I liked AOL up until I discovered that they are very much for the big logger. We pay personal property tax on the skidder. Farmers do not pay any personal property tax on any of their farm equipment. Anyway, a few years ago the association decided they were going to exempt it. Right up till they passed it, I thought they meant this was for everybody ... This took the rug right out from under our feet. It only helps the big companies, who can afford new equipment.[51]

While "big timber" (i.e., corporations) should not be confused with timber workers themselves, these large operators play an important role in keeping the issue of logging in the public eye and political realm. In the Rogue Valley, for example, SOTIA has publicly advocated for harvesting on federal lands, and the media also frequently solicits SOTIA officials for their opinions about forest management.

Unemployed timber workers have also received government support through initiatives like the Northwest Economic Adjustment Initiative's Jobs in the Woods program, which provided assistance and job retraining primarily for displaced timber workers, communities, and others who were adversely affected by reductions in the size of the timber program.[52] Examples of assistance for timber workers in the Pacific Northwest include federal government support, in 1998, of $3.8 million in educational assistance to 875 wood products workers in five Washington counties and the Washington State legislature's passage of a worker-retraining bill.[53] In southern Oregon, the Jobs in the Woods program served both displaced Anglo and Hispanic forest workers.[54] Although such programs have only achieved moderate levels of success in terms of worker retraining and satisfaction, the very fact that federal and state funding was made available for retraining suggests that loggers were a visible constituency to policy makers.[55]

Tree Planters and Pineros

In contrast to loggers, contract tree planters (whether Latino immigrants or US citizens) have enjoyed much less political visibility, although the same cannot be said for all actors in the early reforestation industry. Recall how the Associated Reforestation Contractors (ARC) and the Northwest Forest Workers' Association (NWFWA) represented the competing interests of reforestation contractors and members of tree-planting co-operatives during the late 1970s and early 1980s. The lobbying efforts of ARC played a significant role in re-establishing the dominance of contract crews in the reforestation industry, despite the concurrent organizing efforts of tree-planting co-operatives to defend their interests via the Northwest Forest Workers' Association.[56] The political savvy and organizational muscle with which Anglo co-operative tree planters were able to manoeuvre in the political realm during the 1970s was possible largely because of the skills and resources they possessed, as well as their secure position as citizens or legal residents of the United States.

In contrast to the active political involvement of reforestation co-operatives and contractors, Latino contractors and forest workers in the Rogue Valley today do not participate in public debates or decisions about forest management. In fact, Latino contractors (who are either permanent residents or US citizens) reveal a distinct reluctance to get involved in politics. Some express frustration with the increasing number of regulations and paperwork necessary to become a contractor, but no one has met with agency officials to discuss alternative solutions or policies.

This perspective is quite different from that of Anglo loggers and environmental activists in southern Oregon today and of Anglo reforestation contractors and tree planters in the 1970s. These groups all have attempted to influence the agenda of forest management through varying degrees of political advocacy, public protests, and litigation. Latino contractors, by contrast, do not see themselves or others in their community as having significant political power. When compared to the episodically active role of Anglo contractors in ARC and co-operative members in the NWFWA, the reluctance of Latino contractors to engage in advocacy demonstrates that race and immigrant status play a role in one's occupational marginality. Thus, white US-citizen reforestation contractors (and co-operative members) during the 1970s and 1980s had, and continue to have, greater political visibility than immigrant Latino forest labour contractors in Oregon today.

Pineros, of course, are far more politically marginal than Latino contractors. These workers have historically not been part of formal organizations that represent their concerns to government agencies or the broader public, and that help them seek recourse against labour violations perpetrated by contractors and foremen. Notes from a conversation with Valentín, a pinero who has been living in Medford since 1998, highlight some of the labour abuses and lack of recourse experienced by Latino forest workers:

> Valentín observes that there is not much opportunity to work up. He would like better pay and better treatment. He says workers are abused or treated like animals by foremen and contractors. They are told to work, to get the job done. He doesn't feel like he has any aspirations because he doesn't speak English ... Only foremen speak with *El Forestal* (Forest Service inspectors). Forest workers get no benefits. There is no compensation for workers who are injured in the woods. Or else, even if they are injured, they continue to work. *No hay beneficios* [there are no benefits] ... Valentín says that a union should defend the rights of workers, but he has never seen a union for workers ... He has never seen notices of workers rights.[57]

Stories like Valentín's are fairly common among immigrant workers today. Forest labour has always been dangerous, but in the past, loggers and tree planters were able to rely on organizations that consistently advocated for their interests. By contrast, Latino forest workers have had few such organizational venues. Since the mid-1990s, a handful of non-profit organizations in the Pacific Northwest have tried to raise the profile of immigrant forest workers and create bridges between these groups and government officials and policy makers. These organizations include the Alliance of Forest Workers and Harvesters, the Jefferson Center for Education and Research, and the Sierra Institute (formerly known as Forest Community Research). By allying with policy makers and academics, working with rural immigrants, and publishing well-researched newsletters, these groups have helped raise the profile of various groups of forest workers in the Pacific Northwest.[58] In January 2007, for example, the Ecosystem Workforce Program (part of the Institute for Sustainable Environment at the University of Oregon) held a forum on forest working conditions. The forum provided an opportunity for government officials and policy makers to learn about forest labour conditions, and the Alliance of Forest Workers and Harvesters played a key role in recruiting pineros from Medford to provide testimony.[59]

At the same time, the limited resources, particularly in terms of staff and funding, of non-profit organizations and of the communities with which they work make the sustained involvement and direct engagement of many forest workers difficult.[60] Because it takes a significant amount of time to establish trust and build relationships with community members, the lack of organizational staffing and the member turnover within communities pose challenges to organizations attempting to work with pineros and other rural immigrants. Meanwhile, better-funded and staffed immigrant and labour organizations have generally not focused on the plight of pineros. Despite having the word *pineros* in its name, the largest farm and forest worker union in Oregon, Pineros, Campesinos Unidos en el Noroeste, primarily organizes farm workers, not forest workers.[61] Likewise, Unete (literally, "join us"), a much smaller but highly visible immigrant rights organization in Medford, works mainly with farm workers in the Rogue Valley.

The 2006 congressional hearings on forest labour conditions notwithstanding, pineros overall have been less visible to policy makers and the broader public than Anglo loggers and tree planters. And while non-logging forest workers as a whole have enjoyed much less visibility than loggers, the marginality of pineros extends even further. The dearth of media coverage of immigrant forest workers and the limited access to resources for non-profits seeking to work with them exacerbate the social marginalization of pineros.

While contract forest workers, whether immigrant or non-immigrant, experience labour marginality in a variety of ways, including a lack of employee benefits and limited, if any, political visibility, there are critical differences with respect to the recourse available to workers. Some may argue that laws regarding just compensation apply to everyone regardless of their legal status. However, Latino immigrants are far less likely than white workers to report labour violations and workplace injuries. The marginalization of pineros is generally distinct from that of other forest workers in the Pacific Northwest. When compared to white US citizens who were in the reforestation industry in the 1970s and early 1980s, pineros – whether US citizens, permanent residents, permitted guest workers, or undocumented workers – are less socially visible and have less influence in the realm of forest management politics.

The Marginalization of Pineros

Pineros' marginality results not only from the contingent nature of forest work, but also from various disadvantages that stem from legal status, dynamics of exploitation within kinship networks, and limited English proficiency. In the Pacific Northwest, and in southern Oregon in particular, the majority of pineros are undocumented, with estimates ranging as high as 90 percent of the reforestation and ecosystem management workforce. The majority of these workers have experienced labour abuses, including untimely pay, not getting paid for all work done, and a lack of compensation for injuries on the job. As we have seen, workers rarely reported on-site work injuries because they fear losing their jobs or being deported, or they are not aware of being entitled to workers' compensation.[62] Moreover, because most workers were recruited through family members or friends who were crew members, foremen, or contractors – and thus part of their shared social and occupational networks – they are grateful to have found work and feel obliged to endure hardships on the job.[63] Certainly, keen competition for a place on reforestation crews and the lack of significant or timely enforcement of labour laws mean that crew foremen could easily replace "boat-rockers" with impunity.

In my interviews, most pineros tacitly accepted the present-day system of labour exploitation by voicing their desires for *papeles* (legal papers) so that they too might become foremen or contractors. Finally, the machismo of forest work may also dampen grievances over harsh working conditions and the lack of political voice to redress them: Mauricio Benavides was not alone when he said that danger and injury were inherent to the job and that one had to be tough in order to survive in the woods. Songs and stories glorifying "logging culture," as well as accounts by native-born forest workers, reflect a similarly stoic attitude towards the perils of forest work.[64] However, the romanticization of forest work has not diminished attempts by loggers and other native-born forest workers to seek redress for grievances through policies and court actions. By contrast, their lack of legal standing prevents most immigrant workers from doing the same.

Even when they have legal status, many pineros remain on the margins of forest politics because they lack English proficiency, encounter racial prejudice, and experience a lack of cultural bridge building on the part of land management agencies. The practice of exclusive communication between agency inspectors on Forest Service contracts and crew foremen, for example, exempts predominantly English-speaking government officials from hearing directly from individual Spanish-speaking workers. The ability to speak only limited English affects other immigrant groups as well. The case of Southeast Asian mushroom harvesters most clearly illustrates how legal immigrants who lack English proficiency remain marginal to forest politics. Matsutake mushroom harvesters in the Deschutes National Forest near Crescent Lake, Oregon, were not informed by the Forest Service of plans to log in productive mushroom-harvesting areas. Harvesters only found out about these plans after the fact, when they chanced upon trees that were already marked for logging. The Forest Service only began to communicate with harvesters *after* non-profits working with non-timber forest workers began to advocate on behalf of the harvesters.[65]

Arguably, the Forest Service's failure to reach out to certain groups may not be intentional. Federal agencies often miss portions of the public when sending out notifications of timber-harvesting and road-construction plans, and each on-the-ground project must be written up and published in the *Federal Register*. Agency officials would rightly note that there are opportunities for commenting and that no comments are excluded. Notice of federal projects are also published in local newspapers, and the administrative appeals process offers another stage for commenting on and objecting to such plans and projects.

Yet the question of which groups have access to forums for commentary remains. Environmental justice scholars have long argued that many low-income and minority populations lack the resources (including time and money) to participate in public hearings, especially if travel is required.[66] There are also opportunity costs that come with such participation: attending public hearings would probably reduce low-income individuals' opportunities to work and earn money. One must be relatively educated about bureaucratic procedures in order to know where and how to access records in the *Federal Register* or how to engage in the administrative appeals process. Finally, many immigrant natural resource workers are itinerant; they do not reside in one place (such as a harvesting site) long enough to receive and read notices in the local newspapers, which are usually only published in English.

Language barriers lead to other hurdles such as lack of confidence to approach government officials. Even when workers are able to communicate with government inspectors and attempt to report labour violations, their concerns are often passed on to other agencies. For instance, one respondent noted that when some workers approached a Forest Service inspector about unpaid wages, they were told to refer their complaint directly to the Department of Labor. This only heightened these workers' confusion about policies and procedures, and they never managed to file their contractor's violation of employment law. In another, more serious, case, Latino brush pickers in Washington were arrested in October 2005 by Immigration and Customs Enforcement officials while applying for harvesting permits at the Forest Service office.[67] This incident completely destroyed the pineros' trust in government officials and is likely to hinder any further attempts at communication or co-operation.

Pineros' vulnerable legal status and limited English proficiency may also contribute to aspects of social marginality, such as Latinos' lack of interaction with environmental activists and the logging community in the Rogue Valley. Public protests over salvage logging in southern Oregon typically attract only white environmentalists and logging communities. Although salvage operations also entail reforestation and ecosystem management, neither pineros nor Latino contractors participate in these dialogues. Futhermore, language barriers certainly affect the amount of media coverage given to pineros in the local paper. For example, without any Spanish speakers on staff, the *Medford Mail Tribune* was severely limited in its ability (or willingness) to provide in-depth coverage on the Latino majority doing manually intensive forest work in southern Oregon.

Latinos as "Cultural Spectacle"

Patterns of social marginalization extend beyond forest workers to the greater Latino population in the Rogue Valley. Despite their growing numbers, Latinos in southern Oregon continue to garner little public support when mobilizing around concerns such as immigrants' rights. This lack of support is partially facilitated by mainstream media representations of Latinos as "cultural spectacle" and as non-threatening and apolitical actors rather than as agents struggling for social justice.[68]

The media primarily acknowledges the presence of Latinos in the valley via the lens of "Hispanic culture" rather than in terms of immigrants' political-economic contributions and history of belonging to the area. Popularly recognized markers of Hispanic culture include Mexican food, music, traditional dance, and costume, all of which receive prominent local newspaper coverage. "Latino issues" that are frequently reported in the *Medford Mail Tribune* include restaurant reviews and holiday festivities. A few examples will suffice to show how such items neglect mentioning any relationship to a legacy of immigrant settlement and only perpetuate stereotypes:

> The décor, seasoned with Norteño music (Mexican polka) is a bit surreal for a typical white guy like me, but it fits. Electric lime green walls sport big and small paintings, like my favorite, a train vignette complete with incongruous agave cactus and Roman aqueduct; and my lunch date's favorite, a large black velvet rendering of an Aztec warrior carrying an unconscious woman. Is he rescuing her or taking her to sacrifice?[69]

> With a focus on authentic Mexican cuisine, Nopales Grill has eschewed the popular combination plate smothered in rice and beans. The restaurant's tranquil atmosphere is a welcome change from loud murals, impossibly colored wooden birds, and mariachi music.[70]

> It may prominently feature grinning skulls, but the Dia de los Muertos, the Day of the Dead, is no Mexican Halloween. As celebrated with three altars set up by the Club Latino at Rogue Community College and many other places, it's a heartfelt honoring of dead loved ones along with a celebration that we, the living, have "cheated death" for another year ... Latino students are 10 percent of the population at RCC's Riverside campus and the altars are part of the club's efforts to honor the Latino culture in a creative and personal way.[71]

A local museum exhibit titled "Latinos in Rogue Valley," held at the Southern Oregon Historical Society in the summer of 2004, also reinforced the notion that Latinos are largely viewed in terms of cultural spectacle. The exhibit prominently featured traditional clothing and dance from various parts of Mexico, accompanied by jewellery and other art. Ironically, there was little mention of the history of Latinos in the Rogue Valley, why they settled in the area, or how they contribute to the region's economy and sense of place. Whether it is colourful restaurants or candied skulls, the media and other local institutions thus perpetuate, and the public consumes, highly selective images of immigrant life, framed primarily in exotic terms.

Moreover, not all representations of Latinos in the Rogue Valley are welcome in public spaces. In April 2005, Latino students from Rogue Community College approached Medford's City Council for permission to erect a historic mural downtown.[72] The mural was to depict images of Latinos working in the area's orchards, forests, and service sector, and was part of students' efforts to illustrate, and thereby gain recognition for, the significant contributions of Latino immigrants to the Rogue Valley. City councillors recommended that the project be referred to the Historic Commission and Arts Committee. While the Arts Committee recommended approving the mural, it faced opposition from the Historic Commission, members of which unanimously denied approval of the mural project on the grounds of its size.[73] Students from Rogue Community College, however, felt that members of the Historic Committee denied their mural because they perceived it as "spoiling" the historic character of downtown Medford. Such contrasts between representations of Mexican culture celebrated by many Anglo residents as public spectacle and the lack of recognition of Latinos' actual and ongoing contributions to the Rogue Valley by institutions like the City Council of Medford and the Southern Oregon Historical Society reveal the contested nature of place and continued struggles over "place making."

As Latinos have come to enjoy increased mobility in and between various spaces in the Rogue Valley, they, and young Latinos in particular, are staking more explicit claims to place. More recent public acts of place-making in the Rogue Valley, in March and April 2006, involved Latino participation in rallies over federal immigration policy. Latino and Latina high school students were at the forefront of organizing these rallies, which brought hundreds of Latinos into the streets. Many of these students constitute a younger generation of Latinos who were born in the United States to undocumented parents. In rallying for their parents' legal status, these students displayed a sense of accountability for the older generations' sacrifices. In one protest, a student from a Medford high school held a sign saying, "We could die in Iraq, but our parents can't even get a driver's license."[74] Another rally participant held a sign indicating his long-term status in the Rogue Valley: "With my hard work I earned a family, a house, and a better life style. But I've never earned respect from US citizens."[75]

The issue of immigration reform has certainly raised the profile of Latinos in the Rogue Valley, yet the continuing portrayal (and consumption) of Latinos in terms of cultural spectacle hinders public recognition of immigrants' integral role in the region's political economy and weakens popular support for immigrants' rights and collective action. Seamless contradictions thus embody everyday practices: Rogue Valley residents display anti-immigrant bumper stickers while patronizing Mexican restaurants staffed by undocumented workers; pears prominently symbolize the region's pastoral sensibilities, but the labour behind fruit production remains invisible; and undocumented youth rally for access to higher education in prominent downtown spaces, but Medford newspapers rarely report on such events, thus maintaining the illusion of public harmony.

5
A Tale of Two Valleys

When Tim Buckley moved to Medford in 1980 and became Hispanic Outreach Coordinator for Jackson County Legal Services, the Rogue Valley had seen few attempts at collective action by agricultural workers. Funded by the County Commissioners a few years prior and supported by local orchardists, the outreach coordinator position had traditionally been a low-key affair and not one characterized by overt advocacy. Buckley remembers:

> I'd only been on the job a couple of weeks and they [local growers] invited me to the Rogue Valley Country Club. There had been someone around for a couple of years before me doing what I did. She'd come out and help the foreman's wife get a driver's license and probably spend twenty visits to do that. And that was a good relationship for the orchardists. They felt that they were paying for this outreach worker ... They didn't mind the money, but they didn't want real help for the Mexicans. They wanted some lady, if they [farm workers] got sick, to go in and play "Ave María" for them with a record player. The real economic impact and the real screw jobs that [were] being done on the Hispanics in the community ... they didn't want a remedy to that.

Buckley's critique of growers' resistance to really "help Mexicans" was expressed in light of his own limited success in organizing and advocating for farm workers in the Rogue Valley during the early 1980s. He makes clear that growers were only interested in supporting outreach employees who would preserve the status quo of farm labour exploitation. In this framework, it was perfectly acceptable to adopt a paternalistic attitude towards labour by, for example, soothing ill workers on an individual basis. They did not, however, tolerate organizing against the structural inequalities between labourers and orchardists. The year 1981 saw the first, and last, significant strike among farm workers in the region. Since then, there has been little collective action on the part of farm workers in the Rogue Valley.

Akin to farm workers' limited collective action in the valley, pineros in southern Oregon have little record of organizing against labour contractors. In addition to the challenges presented by their undocumented status, embeddedness within exploitative kinship networks, and language barriers, the migratory nature of field and forest labour hinders organizing in these sectors. The unique isolation of tree planting makes it particularly challenging to organize pineros. Unlike agricultural labour, tree planting usually takes place miles from any passable road or nearby community. It is thus difficult for outside mobilizers to reach pineros while on the job. Because they are responsible for hundreds of saplings that must be planted over large areas, pineros also have limited opportunities to communicate with each other while at work. This, too, is in sharp contrast to farm workers, who labour close to one another when row planting or picking. Finally, there has been an organizational void: no single institution has advocated for or organized around labour rights for tree planters on a sustained basis. Recall how one pinero noted that despite rampant labour abuse, he had never heard of a union that represented tree planters. This sentiment, echoed by numerous others on multiple occasions, reveals the limited reach and visibility of organizations like Pineros, Campesinos, Unidos en el Noroeste (PCUN) and the Alliance of Forest Workers and Harvesters in the Rogue Valley.

Such barriers, however, do not necessarily sound a death knell to organizing immigrant workers as a whole. There have been significant instances of collective action by rural immigrants in similar positions. Consider the familiar historic images of United Farm Workers (UFW) organizers in the 1960s, wielding megaphones on the side of the road and passionately calling on California farm workers to leave the fields, or of pickers circulating pamphlets emblazoned with the black eagle among their compatriots at work.[1] Similar actions are nearly impossible in the forest industry, given the spatial context of the work.

One does not even have to look as far as California's San Joaquin Valley for instances of collective action by immigrant farm workers. Chicanos, Latino immigrants, and migrant workers have a decades-long history of labour activism in Oregon's Willamette Valley, in the northern part of the state. Throughout the 1970s, organizations such as the Valley

Migrant League and Willamette Valley Immigration Project built a legacy of activism and advocacy that later paved the way for PCUN.

The cases of PCUN and the UFW raise the important question of why advocacy-oriented immigrant organizations have been successful in some rural areas but not in others. At first blush, one might be inclined to think that there is something unique about California that makes it ripe for organizing farm workers. Ever since the rapid growth of African American and Latino populations after World War II, California has experienced significant social movements related to race, poverty, and the Vietnam War. Even today, the state is notable for its renewed labour activism among immigrants.[2]

Yet the success of PCUN in organizing farm workers in northern Oregon reveals that immigrant labour activism is not limited to the Golden State. While it may be tempting to attribute the weakness of immigrant activism in the Rogue Valley to limited political opportunities at the state level, one must also take into account variations in labour organizing within Oregon itself. In short, the state context is not a limiting factor. A comparison of the UFW and PCUN with immigrant-serving organizations in the Rogue Valley reveals that historical contexts of political opportunities can be just as critical as contemporary circumstances. In comparing these various cases, we must thus consider not only variation in geography and occupation, but also time periods.

Various organizations have emerged over time and space to address immigrant farm and forest worker issues in the American West. In Table 5.1, I lay out how these cases vary along several dimensions, including location, period of active organizing, and the legal status of members. These organizations also differ in the extent to which they serve as effective advocates for worker rights. As we shall see shortly, the UFW and PCUN rank relatively high in terms of worker advocacy, while the remaining groups score low in this regard. Moreover, the outcome being considered – the level of worker advocacy – does not always correlate with organizational capacity. For example, La Clinica del Valle in the Rogue Valley has considerable organizational strength, particularly with respect to providing affordable health care to low-income populations, including forest workers. At the same time, however, La Clinica has not made a significant mark in terms of mobilizing pineros around labour rights.

So why have some institutions succeeded in organizing immigrants while others have not? One of the key factors that emerges from this comparison is that the era in which groups initially organized remains

Table 5.1

Organizations addressing farm- and forest-worker issues in the US West

Organization name	Location	Period of strong organization	Political orientation	Resistance by owners of capital	Beneficiaries/ members*	Worker advocacy
United Farm Workers (UFW)	Central Valley, California	1967-72	Activist; labour organizing	Yes	Farm workers	Strong
Pineros y Campesinos Unidos del Noroeste (PCUN)	Willamette Valley, Northern Oregon	1977-85 (as Willamette Valley Immigration Project); 1985-present (as PCUN)	Activist; labour organizing	Yes	Forest and farm workers	Strong
La Clinica Azteca	Rogue Valley, Southern Oregon	1986-87	Activist on worker health issues	Yes	Farm workers	Weak
La Clinica del Valle	Rogue Valley, Southern Oregon	1988-present	Accommodationist; collaborates with growers to improve worker health	No	Low-income residents (Latino and Anglo)	Weak
El Convenio de Raíces Mexicanas	Rogue Valley, Southern Oregon	1990-97	Activist on worker health; attempts to form tree-planting co-operative	Yes	Forest and farm workers	Weak
El Centro Hispano	Rogue Valley, Southern Oregon	Weak from founding in 1998-present	Provides interpretation and translation services; no political involvement	No	Latinos with language assistance needs	Weak

* Beneficiaries include both documented and undocumented individuals.

consequential today. Both the UFW and the Willamette Valley Immigration Project, which later became PCUN, were founded and successfully mobilized immigrant workers prior to the 1980s. By contrast, organizations that serve immigrants in the Rogue Valley and that, overall, have been far less radical were all established after 1985. This pattern suggests that there were windows of opportunity prior to the 1980s that made organizing among rural, immigrant farm workers more likely than in the years that followed.

Social movement scholars argue that the timing of protest movements matters in at least two ways. First, timing matters in terms of the windows of opportunity afforded to insurgents by a shifting institutional structure and the ideological disposition of those in power. This perspective looks to elites and argues, using a resource-mobilization model, that a social movement's success or failure depends primarily on external factors rather than on characteristics inherent to the movement itself.[3] Second, timing can matter in that earlier movements may get the lion's share of attention, whereas latecomers in a reform protest cycle may not necessarily confront a weaker state or a more receptive public.

Although earlier waves of Latino immigrants faced many of the same challenges to organizing as pineros do today, they were able to mobilize due to a social and political climate sympathetic to their plight. During the late 1960s, for instance, farm workers and other people of colour successfully harnessed elite resources and spoke out against racial and labour oppression in ways that resonated with the era's national civil rights movement and international struggles for decolonization. In the United States, the UFW was not the only minority group to mobilize against oppression. They were part of a larger anti-establishment wave that included the American Indian Movement, the Chicano Brown Berets, the African American Black Panthers, and student activism against the Vietnam War. The extent to which minorities successfully agitated for and partially achieved social equality during this period reveals a progressive political climate that was at least somewhat receptive to their demands. Significant legislative victories during this period include the passage of the Civil Rights Act in 1964, the historic repealing of ethnic and racial quotas from the Immigration and Naturalization Act of 1965, and the California Agricultural Labor Relations Act of 1975, which allowed workers to vote by secret ballot to choose a union and stipulated that growers negotiate with only worker-elected unions.

Beginning in the 1980s, neo-liberal economic policies took hold both nationally and internationally, generally favouring corporate power and the deregulation of markets over the preservation of social safety nets and labour rights.[4] In the United States, the Reagan administration took a number of steps that hastened the decline of labour. Among the most notable was the firing of over eleven thousand striking air-traffic control workers in 1981. As a result, the air-traffic controllers union was decertified, initiating a long period of decline for both US labour and social equality.[5] The administration also cut taxes for the wealthy, slashed spending on various social programs and aid to states, and removed regulatory obstacles to many US industries, initiating a pattern that has continued with subsequent administrations. Overall, the regressive socio-economic policies of the 1980s and beyond made it difficult for new social justice organizations to emerge and limited the resources of existing organizations.

Earlier gains made by immigrant-serving organizations were not entirely lost during and after the Reagan administration. For instance, the activities of the Willamette Valley Immigration Project during the 1970s laid the groundwork for PCUN to rise as a significant social actor advocating for the rights of farm workers. In the Rogue Valley, by contrast, there was no legacy of immigrant activism to provide a foundation for the Latino-serving organizations that emerged in the mid-1980s. Without such a legacy, the opportunities for advocacy organizations were limited, despite the growing presence of Latino farm workers and forest workers in the region. More generally, what the ensuing analysis suggests is that historical antecedents of labour organizing matter for subsequent efforts, with prior success enabling organizations to survive even in hostile political environments. Absent such antecedents, it is difficult for new activist organizations to take root, especially in prevailing contexts of hierarchical labour relations, increasing corporate power, and the subcontracting of work and services.[6]

The United Farm Workers

Cesar Chavez has come to represent the face and spirit of farm worker organizing in the United States. Born in Yuma, Arizona, during the Great Depression, Chavez and his family moved to California in 1937 to work in the fields, eventually settling in Delano. Chavez cut his teeth as an organizer while working for the Community Service Organization (CSO), founded by Fred Ross in 1948. The CSO worked in Mexican American neighbourhoods to help individuals with a host of issues ranging from filling out tax and immigration forms to gaining access to education. Inspired by CSO's one-on-one model of community mobilization,

Chavez and other former CSO members, including Dolores Huerta, started to reach out to farm workers throughout California. Between 1962 and 1965, they spoke with migrant workers on an individual basis and encouraged them to join their newly founded group, the National Farm Workers Association (NFWA).

Although the NFWA did not initially identify itself as a union, it would soon take on that mantle. In 1965, the Agricultural Workers Organizing Committee, composed primarily of Filipino farm workers, struck against grape growers in Delano, California. Workers sought higher wages and better labour conditions, including access to toilets in the fields and clean drinking water. Shortly after the strike began, members of the NFWA voted to join in the effort. One year later, in 1966, the Agricultural Workers Organizing Committee and NFWA merged to form the United Farm Workers Organizing Committee (UFWOC).

Between 1966 and 1969, Chavez led the UFWOC in spearheading a national boycott on table grapes. He appealed to millions of consumers to support the UFWOC boycott, engaged in non-violent protest tactics, staged high-profile marches, and went on hunger strikes. At its height, the boycott was supported by over 13 million Americans and put enormous pressure on table grape growers, who finally capitulated and signed historic contracts with farm workers in 1969.[7] In 1972, UFWOC was accepted into the American Federation of Labor and Congress of Industrial Organizations and formally changed its name to the United Farm Workers Union (UFW).

There are various perspectives on why the UFW was successful. In an early study on social insurgency, Craig Jenkins and Charles Perrow argued that the UFW's success was due to external resources and "sustained and widespread outside support coupled with the neutrality and/or tolerance from the national political elite."[8] Certainly, UFW strikes and the boycott on table grapes and other non-union produce would not have gone far without the commitment and concrete resources (time and money) of a broad public coalition of allies. The liberal-labour reform coalition included organized labour; religious congregations (the Catholic church played a prominent role in mobilizing public support); politicians who advocated for migrant workers, such as Robert Kennedy, George McGovern, and James Mitchell (who served as Secretary of Labor during the second Eisenhower administration); and the university student population. These allies worked to sustain public attention on farm workers' right to organize and achieve improved labour conditions. Jenkins and Perrow highlight the critical and coordinated role the public played in forcing growers to negotiate with the UFW:

Given the failure of strike actions, a successful outcome required indirect means of exercising power against growers ... To exercise that pressure, a combination of external resources had to be mobilized. Students had to contribute time to picketing grocery stores and shipping terminals; Catholic churches and labor unions had to donate office space for boycott houses; Railway Union members had to identify "scab" shipments for boycott pickets; Teamsters had to refuse to handle "hot cargo"; Butchers' union members had to call sympathy strikes when grocery managers continued to stock "scab" products; political candidates and elected officials had to endorse the boycott.[9]

The federal government, a traditional backer of grower preferences, also remained neutral on labour-related policies due to schisms within the higher ranks of the Republican Party. By 1963, the incumbent Kennedy administration was "in pursuit" of poverty as a public issue and courting minority voters.[10] In hearings about the Bracero Program's renewal, the White House for the first time formally went on record against the program, which was officially terminated the following year.

More recently, scholars have attributed UFW victories to organizational strategy and movement leaders.[11] In this view, success depends on strategic capacity and indigenous organizational strength – a combination of the "leader's life experience, access to salient information, networks, and repertoires of collective action and the deliberative processes, resource flows, and accountability structures within their organizations."[12] In the strategic-capacity framework, UFW success is attributed to Cesar Chavez's leadership role and personal background. Coming from the ranks of farm workers, Chavez effectively harnessed personal experience, culturally and linguistically connected with other workers, and lived among members of the movement. In addition, the UFW organizational structure was not intensely hierarchical. The organization welcomed volunteer support from students and religious figures, and thus relied on a variety of people's dedication and moral authority rather than only on the limited financial and human resources internal to the UFW. Finally, the concept of "cognitive liberation," whereby an aggrieved population becomes conscious of a "problem" and begins to actively mobilize for change, has been linked to UFW success. The use of Catholic imagery and ritual within the framework of liberation theology's "preferential option for the poor," for example, helped build solidarity among farm workers, spread shared understandings about social injustice, and empowered workers to confront perceived oppression.[13]

Regardless of the explanation – whether due to external resources and shifts in political opportunities or to leadership savvy, strategic capacity, and movement participants' cognitive liberation (most likely all) – the UFW's success in organizing workers between 1965 and 1972 is largely undisputed. Farm workers in California were able to exercise their new-found power via ranch committees set up under negotiated contracts and to ensure improved wages and working conditions. By the early 1970s, the UFW had signed more than one hundred contracts, increased wages for its workers by almost a third, and established union hiring halls in every major agricultural area in California.[14] Despite the union's decline in the following decades and more recent criticisms of its operations, the UFW continues to remain a significant player in state and national political circles.[15]

The Willamette Valley: The Legacy of Activist Organizations

The success of farm workers in California coincided with Chicano nationalism and the rise of "brown power." Referred to as *El Movimiento*, this activism was also in full swing in Oregon during the 1960s and early 1970s.[16] During this period, the Willamette Valley became home to several immigrant advocacy organizations that were critical in paving the way for PCUN almost two decades later. These institutions provided services to a critical mass of Latino immigrants who had settled in and around Hispanic-majority towns like Woodburn and Independence, and inspired a new generation of leaders.

Historians Erasmo Gamboa and Carolyn Buan have documented the role of some of these key institutions, including the Valley Migrant League (VML) and the Colegio Cesar Chavez. The VML, founded in 1964, marshalled a coalition of clergy, legislators, Mexican American residents, and agricultural growers to provide a broad range of services to migrant farm workers. Funded through a federal grant, the VML soon developed into a community-based organization that offered adult education, day care, summer school, and health services to migrants and their families. Given federal funding restrictions, the VML maintained a service-oriented rather than activist approach to the immigrant community. Yet the organization also became a platform from which numerous individual employees launched into more direct activism on farm worker unionization. Similarly, the Colegio Cesar Chavez – open between 1973 and 1985 and the only "independent, accredited, and degree-granting institution for Chicanos in the country" – nurtured the development of student activists who later went on to spearhead struggles for immigrants' rights.[17] These institutions developed trust within the Mexican immigrant community in the Willamette Valley through years of community organizing, service provision, and advocacy. More importantly, they sowed a legacy of immigrant leadership and activism from which future organizations, like PCUN, emerged.

PCUN itself was formed from the Willamette Valley Immigration Project, an immigrant advocacy organization started in 1977. Over an eight-year period, the Willamette Valley Immigration Project provided legal representation to undocumented immigrants, sponsored cultural and political events to bring immigrants together, and established a reputation for trustworthiness among both migrant workers and settled residents.[18] Emerging from the Willamette Valley Immigration Project in 1985, PCUN harnessed its predecessor's well-established connections to the immigrant community and grew rapidly. The union's membership expanded from two hundred people in 1986 to over two thousand just eighteen months later.[19] Today, PCUN claims a membership of over five thousand and has active campaigns on immigration reform, access to higher education for the children of undocumented workers, and protection for farm workers from pesticide exposure. Not incidentally, PCUN's founding members included Ramon Ramirez, Cipriano Ferrel, and Juan Mendoza, all of whom had cut their teeth as student organizers at the Colegio Cesar Chavez.

The point here is not to detail the various organizing challenges and successes faced by PCUN over the years. A number of scholars have already written about the union and its role in both organizing farm workers in the Willamette Valley and challenging anti-immigrant and anti-farm worker legislation more broadly.[20] What is important is that PCUN's viability depended on the work of predecessor institutions in the Willamette Valley. The social activism of groups such as the VML, Colegio Cesar Chavez, and Willamette Valley Immigration Project were foundational to the success of later immigrant advocacy in the region. These early institutions built links to immigrants, grew community leaders, and lay the groundwork for future organizing. PCUN was not formed in a vacuum; it harnessed the reins of activism around immigrants' rights that had been in the works for decades. Thus, despite the conservative turn in politics during the Reagan era, PCUN still successfully emerged in the mid-1980s. Immigrant organizing in the Rogue Valley was not as fortunate. With no legacy of earlier waves of immigrant advocacy in the area, organizing efforts in the region either failed or were co-opted into less transformative models of change.

The Rogue Valley

The Farm Worker Strike of 1981

The United Farm Workers' sustained and successful work stoppages and table grape boycott during the late 1960s could not have been more different from the short-lived strike among farm workers in the Rogue Valley in 1981. That year, a small group of individuals, including Tim Buckley, gathered a petition with four hundred signatures demanding a raise in the bin price of pears. Buckley remembers:

I wrote out this petition to the Fruit Growers League and they of course wanted to see it. Nineteen eighty one was going to be a big harvest year, a huge crop. And they, of course, didn't want the apple cart upset. I wouldn't turn the petition over to them because I didn't want them to see the names on it. But they accepted the fact that there were over four hundred Mexicans out there on their crews that signed this petition.

What they were asking for was a raise in the bin price. These bins are huge. And they were paying them $7 a bin. And every year they would raise that by a quarter, we are talking about a quarter of a dollar. Well, we are talking about [a] 3 percent [raise]. [This is during] 1977, '78, '79, Jimmy Carter era when inflation was between 12 and 16 percent. So 25 cents is nothing, they were not raising the wages at all. I've picked apples before. Apples are light and they are big. But pears are small, they are heavy. The straps put grooves in your neck. Pear picking is a hard job and so to pay these guys an extra quarter a year, which would have brought them to $8 that year, was nothing. Most of these guys would do about three bins a day when they started and then move up to between six and eight a day. But we are talking about supermen here, the cream of the crop from Mexico ... young bucks who are out there, not drinking, just working.

So in this petition, we asked them, the growers, to raise their bin price from $8 to $10. *Now that's radical.* That's a whole lot of money. That's a 20 percent [wage] increase, that's big time. Now, the growers knew what they were up against, they had been dealing with me for a year already. And they actually somewhat did it. They raised the price a dollar, from $8 to $9, which was huge, and then they put in a dollar bonus for each bin if you finished the season. But of course, that bonus was voluntary.[21]

Specifically, farm workers demanded a $2 or $3 wage increase in the bin price of pears: from $8 to $10 for Bartletts, $9 to $11 for Boscs, and

$9 to $12 for Comice and d'Anjou pears.[22] As Buckley notes, workers' wages had not kept pace with high rates of inflation, and their demand for a raise was long overdue. Indeed, the per-bin minimum price for Comice and Bosc pears had not been increased since 1975, and the wages for Bartlett pears had increased by only $1 that year.[23] Newspaper accounts indicate that a strike *threat* by 175 Hispanic farm workers in 1975 may have led to the dollar increase in wages for Bartletts. However, orchardists were able to get pickers back to work and thwart any potential strike action. As one grower remembers of the organizing attempt in 1975:

> And, of course, we have not had a great deal of labor problems from pickers and so forth ... Several years ago there was an effort to picket. I was coming out from town and a bunch of them, mostly Mexicans, they'd learned [about the picket] and cars all over the place, they just parked out there where we turn off onto Modoc Road, it's narrow and there's no place to park, and the State Police were out there. A lot of them had pulled off the road and there was just barely room to get through, they'd parked on both sides of the road, strictly illegal and all that, but the State Police weren't doing anything about it even though they were impeding traffic. But I thought, "Uh oh, heaven knows what this is going to work into," but apparently nothing. That was the only incident. I did hear they had done this a few times elsewhere. But apparently they didn't make that stick either. I guess the fruit workers do negotiate though. I don't know what it is, with the fruit growers. I know they're sure paying a lot higher wages than they used to.[24]

The strike in 1981 was the first long-term work stoppage the Rogue Valley had witnessed in decades and was directly related to farm workers' demand for higher wages. While most orchardists agreed to a $1 wage raise – which in itself did not meet workers' initial call for a $2 increase – not all agreed to the $1 per bin bonus. It was in this context that eighty farm workers at Pinnacle orchards decided to strike. The Pinnacle strike received state-wide media coverage and lasted just under a month, but three weeks into the strike, only eight strikers remained. Ultimately, growers' refusal to comply with the demand for a retroactive bonus prevailed. The inability of most workers to "stick it out" – most found work in other orchards or left the area altogether – highlights just how difficult it was to sustain a prolonged strike and *win*. Buckley recollects: "It [the strike] was complicated. You are talking about eighty people, this is their livelihood. This isn't just something to just have fun with and joke around with. You are talking about real people with their futures at stake ... In the end, seventy-two took off. Only eight stayed behind ... The company wouldn't bend."[25]

In contrast to the UFW, farm workers in the Rogue Valley had no union to pay their wages when they struck or to help them collectively bargain with growers. They also received little public support or sympathy for their demands. Buckley further noted that even members of the Latino community refused to overtly support the strikers. When Buckley asked several community leaders to sponsor a *baile* (dance) to raise funds for and rally behind the farm workers, they refused so as not to antagonize valley growers. This is in sharp contrast to the widespread support that the UFW received through the late 1960s. There was also no clear strategy or leadership behind the 1981 strike. In contrast to the highly visible campaign spearheaded by Cesar Chavez and backed by numerous organizers throughout the state, the Rogue Valley strike was limited to one labour camp with fewer than one hundred workers. With few resources to support the strikers and little organizational leverage to bargain with orchardists, the majority of strikers eventually moved on to find other work.

While the 1981 strike was unsuccessful, it did lead to marginal changes in the negotiation strategies of some growers. The following year, for instance, when workers in Rogue River threatened to strike, several orchardists were willing to negotiate, albeit on their own terms. Overall, however, there has been little sustained collective action on the part of farm workers in the Rogue Valley. Moreover, this lack of coordination among farm workers can partially be attributed to a lack of institutional support. Without a union or other advocates to back them, farm workers were left to fend for themselves as small and uncoordinated groups. The fact is that there have been few immigrant advocacy organizations in the Rogue Valley.

From Immigrant Advocacy to Service Provision

Unlike the UFW in California and advocacy-oriented institutions in the Willamette Valley (WVIP, VML, PCUN), organizations serving immigrants in the Rogue Valley have mainly concerned themselves with the limited provision of social services. Attempts at immigrant organizing have generally been short-lived. Why have advocacy groups similar to the UFW failed to take root in southern Oregon? Why are politics not more radical in places such as the Rogue Valley? And how does this bode for immigrants' rights and well-being in the long run? Examining the unsuccessful attempts to mobilize farm workers in the Rogue Valley by La Clinica Azteca and El Convenio de Raíces Mexicanas helps to answer these questions. La Clinica Azteca was a community health clinic founded by and for migrant farm workers in 1986. In its short life, Azteca openly challenged Rogue Valley growers over pesticide use and farm worker health. El Convenio, too, was a community organization started by farm workers. It took on pesticide issues and also attempted to create avenues for immigrants' autonomy via a community garden and tree planters' co-operative. In both cases, the activist elements within these organizations faced strong opposition from existing power structures and were eventually purged.

By "activist elements," I mean organizational actions or agendas that seek to confront power imbalances and/or remedy injustices faced by members of the Latino community. The UFW and PCUN, for example, are radical in the sense that unionization enables farm workers, a historically powerless group, to negotiate with growers and improve their working conditions and wages. Organizations such as the WVIP and VML, which were largely oriented towards service provision and were prevented by federal regulations from taking on an overtly political role, were nevertheless pressured by individual employees to provide more support to Oregon's farm workers. Gamboa notes:

> This pressure only worked to strengthen VML's commitment to Oregon farm workers. Often working behind the scenes, the organization did much to alter their lives in a meaningful way. Because of its concern with health conditions in farm labor camps and the migratory cycle of Oregon farm workers, the VML developed a federally funded self-help housing project called Farm Worker Housing that allowed farm workers to build and own their homes ... Through self-help housing, the Valley Migrant League – and later Farmworker Housing Project – addressed problems that had persisted for decades, problems related to education, health, and temporary residency.[26]

The "brown power" commitment of employees within service organizations thus pushed their institutions to support a broader immigrants' rights agenda during the 1970s. The VML's success in attaining federal funds for a farm worker housing project also illustrates sympathies for immigrants' rights at the national level. By the late 1980s, this climate had changed. Despite the passage of the Immigration Reform and Control Act and the related amnesty in 1986, there was little overt support for confrontational politics on the part of service-provision organizations in the Rogue Valley. For example, La Clinica Azteca staff actively opposed actions that might inconvenience the area's growers. As a farm worker organization, El Convenio also faced an uphill battle in securing funds for farm worker housing and more generally gaining the trust of local government officials. Ultimately, both of these institutions closed down. A closer look at the histories of La Clinica Azteca and El Convenio uncovers some of the dynamics at play.

From La Clinica Azteca to La Clinica del Valle

In 1986, La Clinica Azteca, a federally funded health care centre for migrant workers, opened in Phoenix, OR, about ten miles south of Medford, the county seat of Jackson County. With a board of directors composed of farm workers and led by a Hispanic woman from the Willamette Valley, La Clinica Azteca lasted only a year before its funds were revoked and the organization was shut down. Azteca's executive director and the farm worker board members were not originally from the area, and they had direct experience working in the fields. By contrast, the staff was primarily made up of white residents from the Rogue Valley, none of whom had a background in agricultural labour.

According to former clinic employees, Azteca's closure was due to both poor management and the executive director's stance against pesticides, which the orchardists in the area found particularly troubling. When asked about orchardists' experience with labour issues, for instance, a long-time member of the Rogue Valley Fruit Growers League referred to Azteca's former director as the "Dragon Lady":

> We had what you call the "Dragon Lady." That was the name we gave her because she was so contrary, I guess you could say, somewhat vicious towards the industry. She worked the labour force up tremendously about how awful things were for them ... We had a meeting with the growers in this very office to try and deal with the problem ... Ever since then, there hasn't been a labour stoppage, except for maybe an isolated case with one grower ... Of course, there is always the rotten apple that doesn't treat his people right. But the Fruit Growers League itself, by charter, is not to enter into labour agreements.[27]

Azteca's staff also did not see eye to eye with the clinic's executive director and expressed their dissatisfaction by taking turns being absent from work, a situation that eventually led to an outside inquiry. As one former employee recalled:

It came to a point where we just decided to strike against the director. We would take turns not coming to work. Finally the feds, the federal government, which we received funding from, came down and saw the situation was in crisis. And they pulled the funds away from Clinica Azteca and said, "We are not going to fund you, you need to get a new board."[28]

Interestingly, organizational mismanagement and the director's anti-pesticide position were often conflated. Whereas the all–farm-worker board stood behind the executive director, Azteca staff did not approve of her confrontational attitude regarding pesticides. Indeed, the staff often expressed more concern over growers' resistance to pesticide politics than over the potentially adverse effect of pesticides for farm workers. Although Azteca's staff may not have directly empathized with growers, it is clear that they wanted to gain legitimacy with this constituency. Furthermore, Azteca employees not only disapproved of their director's confrontational politics – an attitude reflected more generally by Rogue Valley residents – but also equated it with incompetence and ignorance. Characterized as outlandish and inappropriate, and linked with poor management and a lack of knowledge, Azteca's radical politics were hamstrung from the outset and eventually led to the clinic's demise. The former employee continued:

The feds came down and met with different people from the Health Department, medical community, and growers – local orchard growers were also very unhappy with the happenings of the original organization around pesticide issues. The original executive director made all sorts of outlandish and inappropriate remarks about pesticides. Showing her lack of knowledge. Anti-pesticide, kind of like we're going to shut the growers down sort of thing. They [feds] came down to talk to physicians and staff, to find out what the problem was. And after speaking with everyone, they were convinced that the original organization was just not going to be able to function. And they also understood and supported the employees and staff leaving the original organization.[29]

After the dismissal of its original director and the dissolution of the board, Azteca reopened as La Clinica del Valle (LCDV or La Clinica) in 1988. The new director, a former Azteca employee, made it her mission to gain the trust of the alienated grower community. LCDV's stance on pesticides and overall policies towards orchardists could not have been more different from those of its predecessor. Rather than confrontation, the new clinic's relationship with growers may be best described as one of accommodation and co-operation.

LCDV staff correctly perceived the power and influence wielded by valley growers and saw them as necessary allies for the long-term success of the clinic. This meant shedding any link to radical farm worker politics and convincing growers that the new clinic would not only avoid causing "problems" but would also help increase their bottom line. In practice, this has meant that La Clinica communicates directly with orchard owners to resolve issues regarding farm worker health. Such an approach means working with management rather than against it. It focuses on maintaining worker health and productivity rather than on increasing worker empowerment and education around labour rights. The conscious strategy of cultivating trust with the grower community was clearly expressed by LCDV's administration:

I mean when LCDV started, we made it very clear that it was equally important to the growers as it was to La Clinica to keep workers healthy. And that we would keep our lines of communication open. And if we had any concerns about people coming in with pesticide issues, that we would work directly with the grower to resolve them. And that took quite a bit of time to get the confidence of the growers that we were ... that we were legit. And that we didn't always agree on issues, but there was enough mutual respect that when issues came up, we were able to work through them with individual growers.

The more we tell the story that we are not this radical non-profit group causing headaches for the growers, [the more financial support] we get from them. That's been the reputation of migrant health centres unfortunately. There has been a history of tensions between farm workers and growers, especially as it relates to pesticides. The fact that we don't have that here, they [growers] invite us to the camps to do our education pieces ... The growers, from an employer perspective, recognize that by keeping their employees healthy, it works better for them. It is better for their bottom line.

I think if the way in which we approached it [pesticide issues] felt radical for them [growers], then they would be touchy. I think if we approach it as "we are providing information, this is health education, it has a direct impact, that you the grower may not necessarily be to blame, that you have good intentions, that you don't go out to intentionally harm people," approaching it that way, it seems to have good success [with the growers] ... A "no-attack" approach.[30]

Since its inception in 1988, LCDV has expanded to serve the growing numbers of underinsured and uninsured in Jackson County. In 2009, 72 percent of La Clinica's patients earned below 100 percent of the federal poverty level and more than 52 percent were uninsured.[31] As such trends increase, LCDV plays a critical role as the only full-time sliding-scale health facility in Jackson County. It serves thousands of valley residents and migrant workers who would not otherwise have access to health care. To meet growing demand, the clinic already has three facilities in the valley. La Clinica's ability to bring in continuing funds is largely due to its positive relationship with, and endorsement by, valley growers. Mike Naumes, the nation's largest pear grower and a prominent Rogue Valley orchardist, led the clinic's capital campaign (which raised $1.3 million) for its West Medford facility. Other prominent members of the grower community, including the president of the Valley Fruit Growers' League, also sat on the campaign's steering committee. Overall, LCDV raised over $1 million in funds locally. More importantly, its ability to do so depends on maintaining standing within the resource-rich grower community – a feat La Clinica achieved through shedding its erstwhile history of confrontational politics. For growers, the clinic has thus gone from being a potential thorn in their side to a trusted institution whose services they not only tolerate, but actively solicit. As one LCDV employee noted:

> The growers have completely opened their camps to La Clinica. They *expect* La Clinica to come to the camps now. It has become, over the years, an expectation. Originally, we went in because we could and they couldn't stop us. And then we came in because they said, "okay, you can come in" and now, they are like "you've *got* to get to all the camps."[32]

Finally, La Clinica has achieved trust and legitimacy within the grower community by working with rather than against orchardists. This has meant that health, safety, and labour-related problems are kept "in-house." In offering services to migrant workers, then, LCDV also implicitly quells labour radicalism and worker empowerment. This outcome becomes possible through the co-operative (some might say co-opted) stance that the clinic takes with orchard managers:

> At La Clinica, we made it really clear that there were services that we could offer the people in their camps that would benefit the growers as well. We could go out and test everybody for tuberculosis, we can talk

about sexually transmitted diseases, we can deliver condoms, follow up on any health problems, sign people up for the Oregon Health Plan. All these different things that would benefit growers themselves. And in return, if any issues came up that were either safety related or pesticide or something like that, then we would go directly to the orchard manager or grower to resolve it, before going elsewhere.[33]

El Convenio to Centro Hispano

Jose Monterrey moved to the Rogue Valley in 1991. He found work in one of the area's large packing houses and shared a ramshackle home with nineteen other farm workers. While on the job, Jose heard about El Convenio de Raíces Mexicanas, an organization started by and advocating for farm workers. Founded in 1990, El Convenio addressed pesticide issues and workers' rights. It also attempted to organize immigrants through a tree-planting co-operative and community garden. By 1994, the organization was over two hundred strong, had an annual budget close to $100,000, and received funding from a number of small social justice foundations. Yet three years later, El Convenio had folded amidst financial mismanagement and the subsequent loss of funding. What happened?

According to former Convenio members, the organization fell apart due to internal factions and external pressure. Internally, conflicts between old and new board members became increasingly visible to the public, as did questions of leadership. Rather than uniting behind El Convenio, various members of the Latino community fought over leadership positions and, as one former director recalls, "wanted all the credit." Accusations of financial mismanagement also pervaded the organization. Although outside investigators found no basis for these accusations, the organization's reputation suffered and foundations discontinued funding. The executive director eventually stepped down and the old board dissolved.

Throughout its brief existence, El Convenio received little support from members of the larger Rogue Valley community. As in the case of La Clinica Azteca, local corporations and fruit growers did not approve of the group's focus on pesticide exposure and farm worker empowerment. Thus, when El Convenio hit financial straits, no local donors were willing to help keep the organization afloat. This is in stark contrast to the fundraising and donor support for La Clinica del Valle's continuing expansion.

What found support, instead, was a new incarnation of El Convenio called El Centro Hispano. Unlike its predecessor, Centro Hispano has taken on a much more traditional, service-oriented role. Operating on a shoestring budget and staffed by one individual on a part-time basis, Centro Hispano currently offers basic interpretation and referral services for the Latino community. It maintains a low profile and its public appearances are limited to hosting information booths at events such as the Medford Multicultural Fair and Farm Worker Day. Most significantly, Centro Hispano is not a social justice organization that questions or challenges issues of immigrant or farm worker marginalization. The status quo of immigrant service provision, without empowerment, has thus generally persevered among organizations in the Rogue Valley.[34]

Explaining the Lack of Activism

My analysis of Latino labour activism in rural areas suggests that the context of initial mobilization plays an important role in shaping the success of future organizations: institutions formed during the 1960s and 1970s left a legacy of strong advocacy for immigrant rights, while those formed after the "Reagan Revolution" were much more constrained in their activism and focused primarily on service provision. My emphasis on the significance of historical context aligns with sociologist Doug McAdam's influential theory of the political process model of social movements.[35] In re-examining the African American civil rights movement of the 1960s, McAdam moves away from social movement theorizing that centres on factors such as relative deprivation (creating pressures for mobilization) and elite sponsorship (outsiders supplying resources for organizations). He argues instead for a political process model, which stresses the importance of political opportunities at the institutional level that provide the overarching context for social change, internally generated resources to capitalize on those opportunities, and cognitive liberation among individuals who recognize a problem and see the need to mobilize.

In my analysis of varying outcomes with respect to immigrant organizing, the timing of initial activism forms part of the political opportunity structures currently available to organizations. The Willamette and Rogue Valleys could not have had more divergent paths when it comes to mobilizing immigrants. Immigrants have successfully spearheaded organizing efforts in the Willamette Valley for over three decades. Their activism during the late 1960s was energized by the broader political climate of the civil rights era, when Chicano and other subaltern movements found allies among politicians and the general public. Support for farm workers' rights included religious organizations, college students, and progressive politicians, all of whom donated financial resources, time, and labour to help with organizing campaigns and supporting boycotts.

The Willamette Valley also became a key site for immigrant activism due to its critical mass of Latino residents. The region saw the settling of ex-braceros and Tejanos starting in the 1950s, followed by Mixtecs and Triqui migrants from southern Mexico in the late 1970s.[36] By the 1980s, places such as Woodburn and Independence were close to being Latino-majority towns and served as headquarters for advocacy organizations such as the Valley Migrant League, Willamette Valley Immigration Project, and Pineros y Campesinos Unidos en el Noroeste. A long-settled and high-density Chicano and Latino immigrant population thus enabled the establishment of advocacy institutions geared towards immigrant rights and services in the Willamette Valley. Many of these organizations began during the civil rights era and explicitly confronted issues of social marginalization and labour exploitation. More importantly, they established a foundation upon which future generations of Latino activists have continued to build.

By contrast, immigrant activism in the Rogue Valley has been sparse and largely unsuccessful. On the one hand, both the farm worker strike in 1981 and La Clinica Azteca's stand against pesticides found little support in the larger community. On the other, more service-oriented and status quo institutions like La Clinica del Valle have prospered. The failure of more explicit immigrant activism may partially be explained by two interrelated factors: the relatively late settlement of Latinos in the Rogue Valley and a lack of precedent for later waves of advocacy.

Latino settlement in the Rogue Valley only began in earnest during the early 1990s, and a large portion of this population is undocumented. Institutions serving immigrants in southern Oregon, unlike their counterparts in the Willamette Valley, neither benefited from a historic legacy of activism nor were founded in a political climate supportive of organizing. With little institutional support and even less public sympathy, compounded by considerable resistance from the Rogue Valley's well-connected orchardists, the nascent organizing efforts of farm workers and their advocates failed to gain traction during and after the 1980s. Thus, whereas support for PCUN grew tremendously after its founding in 1985, organizations such as La Clinica Azteca were publicly derided for their attempts at activism and swiftly shut down.

While the legacies of prior mobilization thus clearly play a role in shaping current political opportunities, other factors may also be important. Within the rubric of political opportunities, proximity to major cities and the isolated nature of forest work may also play a role.

Indigenous resources such as leadership skills and organizational capacity may further help explain the variation between these cases. Below, I consider these explanations, showing the strengths and limitations of each in accounting for why immigrant worker mobilization has succeeded in some cases but not others.

The Nature of Forest Work

For a number of reasons, mobilizing forest workers is more challenging than organizing farm workers. Farm labourers pick and plant in close proximity to one another, adjacent fields are nearby, and workers and work sites are accessible to organizers by road. By contrast, pineros work in locations that are geographically isolated and inaccessible by road, and workers' ability to communicate is severely hindered by their spatial separation while planting. Moreover, even when pineros are in close contact, such as on their long rides back home, they are limited in their ability to discuss organizing given their sheer physical exhaustion and the near-constant presence of management (foremen or contractors who speak Spanish). The nature of forest work may thus constrain both the opportunities and resources necessary for successful worker activism.

While the argument that it is nearly impossible to organize forest workers certainly has merit, two points must be considered. First, the case of tree-planting co-operatives like the Hoedads in the late 1970s and early 1980s shows that there is some historical precedent for forest worker activism. As I detailed in Chapter 2, the Hoedads were a fairly strong organization that won contracts on federal land, set planting standards, advocated for worker interests, and, for a time, competed effectively against contractors and contractor organizations. Second, the inclusion of pineros in PCUN indicates that there is some contemporary variation in forest worker activism, with pineros in the Willamette Valley having much greater access to worker advocacy than do those living in the Rogue Valley.[37] At the same time, these two counterexamples also highlight the limits of organizing pineros. The case of the Hoedads (which pertains largely to white, US-born workers) reveals that forest worker activism may wither in the face of political and economic competition from contractors. Also, the case of PCUN shows that even those organizations that attempt to organize both pineros and farm workers have tended to advocate much more frequently on behalf of farm workers than forest workers.

Still, these two examples offer an important corrective to the notion that it is impossible to organize forest workers. In the case of PCUN in particular, it is difficult to disentangle the explanations of occupational disadvantage and legacy effects of prior activism. While it may be tempting to explain the greater advocacy on behalf of farm workers as based on occupational differences between farming and forestry, one can also argue that these differences are the result of legacy effects: PCUN's focus on farm workers has a much longer history in the organization and is directly related to the work of predecessor organizations such as the UFW and Colegio Cesar Chavez. Thus, while there is some merit to the argument that organizing forest workers is more difficult than organizing farm workers, the cases of the Hoedads and PCUN show that organizing pineros is not an entirely impossible endeavour.

Geographic Isolation

As Mexican immigrants settle in new destinations across the United States, a growing body of scholarship suggests that geographic isolation from large metropolitan areas can be disadvantageous to immigrant civic organizing. For instance, the density of immigrant-serving non-profits tends to be much lower in rural and suburban areas than in central cities. This, in turn, makes it more difficult for immigrants in less densely populated areas to gain access to social services, government resources, and language assistance.[38] The presence of unions and charitable foundations is also more sparse in rural areas than in urban areas, further hindering the ability of new organizations to take root.[39] The differences in organizing immigrant workers may thus also be due to political opportunities and resources related to geographic isolation.

In making the comparison between the Willamette and Rogue Valleys, such an explanation certainly seems plausible. PCUN's centre in Woodburn, Oregon, is thirty miles south of Portland, a city of over half a million, while the population centre of the Rogue Valley is Medford, a city of about 75,000 residents. One might thus expect immigrant workers and immigrant-serving organizations in the Willamette Valley to have much greater access to elected officials, community foundations, and political allies than do those in the Rogue Valley, and the cases do indeed bear this out.

However, as the case of the UFW in Salinas, California, indicates, geographic isolation is not an insurmountable obstacle to organizing immigrant workers. Even though Salinas was nearly two hundred miles away from Sacramento and over three hundred miles from Los Angeles, Cesar Chavez was able to build alliances with labour organizations and other liberal groups in various parts of the state to advance the interests of farm workers in the central valley and central coast of California. By contrast, leaders of immigrant-serving groups in the Rogue Valley have not succeeded in creating similar alliances with PCUN or other labour organizations in Oregon. Indeed, Unete is the only organization in the Rogue Valley that has sought to partner with PCUN, and it remains marginal to the civic life of immigrants in Medford: it is essentially a one-person outfit that is seen as too radical by most local service providers, including immigrant-serving organizations.

More generally, the literature on immigrant civic organizing indicates several ways to overcome the disadvantages of geographic isolation. As Kristi Andersen has noted in her study of immigrant communities in places like Chico, California, and Fort Collins, Colorado, public universities can serve as important allies, with volunteer programs, language assistance, and even worker advocacy.[40] Immigrant leaders can also rely on churches and social service providers as important sources of organizational support as they build up the civic capacity of their communities. Finally, time can also help to overcome barriers related to geographic isolation, as areas with a longer history of immigrant settlement are more likely to draw on allies from established institutions such as churches and schools. Such institutions have certainly been present in the Rogue Valley: Southern Oregon University has two campuses in the region with over two hundred academic staff and Rogue Community College has over five hundred instructors in the region who could serve as potential allies for immigrant-serving organizations. However, while the university community has actively engaged with the concerns of environmentalists (by hosting conferences and providing student volunteers), the same has not been true regarding advocacy around immigrant labour. Given that Latinos have been living in the region since the early 1990s, this lack of institutional outreach is not attributable to the recency of immigration. Rather, the Rogue Valley lacks a mobilizing culture that is supportive of immigrant rights – an absence partly attributable to the absence of civil rights activism during the 1960s.

Leadership and Organizational Capacity

Leadership capacity may also account for the variations in worker advocacy between PCUN and Rogue Valley institutions. As noted, PCUN's founders cut their organizational teeth in the social and racial justice politics of the 1960s. Individuals such as Ramon Ramirez and Cipriano Ferrel were passionate public speakers and experienced popular educators with deep ties to the communities that they mobilized. Thus, PCUN was focused on a vision of worker advocacy and empowerment from the outset and was led by individuals who brought skills, charisma, and legitimacy to this project. By contrast, the failure of La Clinica Azteca's director to mobilize farm workers in the Rogue Valley might signal a lack of leadership capacity. Azteca employees, for instance, noted the unprofessionalism, disorganization, and conflict orientation of their boss, who was eventually fired.

Yet it is likely that even a more professional director with better organizational skills would have failed if she had maintained a radical political stance. Recall that Azteca employees tended to conflate "unprofessionalism" with particular political viewpoints. More specifically, a strong anti-pesticide stand and open confrontation with valley growers led to accusations of "poor leadership" and Azteca's early demise. By contrast, La Clinica del Valle, Azteca's successor, has thrived in its role of providing health services to immigrant workers in the Rogue Valley. While this is partially a testament to LCDV's leadership capacity, it is just as importantly due to the management's explicit efforts to avoid confrontation.

Such dramatic divergences in the fate of Rogue Valley organizations reveal that outcomes are not simply a result of differences in leadership capacity or resources. Rather, local political culture – and the underlying dynamics of social power (who holds power, who controls resources, who makes decisions, what groups are influential in promoting or quashing agendas, etc.) that shape it – play a central role in determining the horizon of possibilities for mobilization. Moreover, "mobilizing culture" is not static. Organizing efforts in earlier eras have an effect in changing the characteristics of an area and building a legacy for future activism.

Immigrant workers in numerous industries continue to face labour exploitation and abysmal working conditions. They currently confront a context of increased workplace raids and punitive, anti-immigrant policies. The work site enforcement of immigration laws has led to the detention and deportation of thousands of undocumented workers, separated parents from their US-born children, and instilled fear and confusion within the Latino community.[41] At the same time, violations of labour law largely go unprosecuted, and immigrant workers remain invisible and unrecognized for their significant economic contributions.

In a national context in which undocumented workers fear legal repercussions (exacerbated by recent anti-immigrant legislation in states like Arizona and Florida), immigrant-serving institutions face significant challenges to organizing. This is not to say that there has been no organizing or resistance. The immigrant marches in Medford and across the country in May 2006 revealed that thousands were willing to step forward and demand an end to living in the shadows. The activism of Latino youth also illustrates the ongoing nature of the immigrants' rights struggle and is a promising, yet limited, hope for reshaping "mobilizing culture" in the Rogue Valley.

Individual workers are also capable of resistance, if not mass mobilization. For example, pineros shirk on the job to avoid exhaustion from faster planting schedules or quit the job altogether. Such acts inconvenience contractors by lowering their crews' overall productivity and ability to meet contract deadlines. Still, to romanticize such instances of individual resistance as effective "weapons of the weak" ignores the significant challenges to large-scale organizing by rural immigrant workers more generally and belies their efficacy.

In the Rogue Valley, pineros and their families have yet to be fully acknowledged for their socio-economic contributions or treated as an integral part of the local community. Legalization programs may help to improve the standing of these workers, but unless there is a large-scale shift in the dominant society's attitudes and perceptions, Mexican immigrants in these areas, illegal or otherwise, will remain peripheral to local policy debates and sense of community. Past and present radical advocacy organizations have played a critical role in establishing precedents for and sustaining movement building and empowerment among immigrants. Their role – or lack thereof, in the case of more pro-status quo groups – in catalyzing shifts in public support and related social policies should not be ignored.

6
Conclusions

Debates over environmental policy in the Pacific Northwest are marked by a series of omissions: a dearth of information about manually intensive, non-logging forest labour on federal lands; the absent history of the Latinization of forest work and settlement of pineros in rural, natural resource communities; and the marginality of Latino immigrants in environmental debates and policy considerations about forest management in the Pacific Northwest.

The extant scholarship has overwhelmingly focused on the timber-related aspects of forestry and the labour of white male loggers. Despite the decline of timber harvesting on federal lands, the myth of Paul Bunyan continues to have cultural resonance today. This inertia of imagination and scholarship deflects attention from the understory of reforestation work. During the early 1970s, Anglo contractors and subcontractors hired a low-wage, mostly white US-citizen labour force to plant trees on federal lands. By 1985, most Anglos had left the reforestation industry to pursue other employment or educational opportunities; they were replaced by immigrant Latino workers who continue to dominate the ranks of manually intensive forest labour today.

As Chapter 3 showed, the gradual Latinization of forest work in the Rogue Valley may be attributed to a number of factors. Early Anglo labour contractors had ready access to a supply of low-wage agricultural workers and thus did not have to go far to recruit immigrant tree planters. After an initial period in which contractors directly recruited Latinos from the fields into the forests, pineros primarily entered forest work through their own social and kinship networks. As a result, tree-planting crews became ethnically homogeneous, composed of Latino workers who were often from the same home village in Mexico and/or related through kinship. Similar labour-recruitment strategies are common to immigrant workers in other sectors (e.g., agriculture and service industries) and reinforce the important role networks play in expanding ethnic enclaves in new rural destinations.

The entry of Latinos into labour contracting was contingent on specific federal laws and policies. The passage of the 1986 Immigration and Reform Control Act and its related amnesty granted legal status to millions of undocumented workers and opened up opportunities in labour contracting. Latino contractors also benefited from the Small Business Administration's set-aside programs and from their access to low-wage labour through social and kinship networks, factors that continue to sustain the dominance of Latino forest labour contractors in Oregon today. However, because the Immigration and Reform Control Act only granted amnesty to a limited pool of applicants, it led to the segmentation of forest labour along the lines of legal status, with newly documented Latino labour contractors primarily hiring undocumented Latino workers.

Pineros today are among the most economically and socially marginal groups of forest workers on federal lands. Chapter 4 demonstrated that pineros' labour marginality is related to their lack of legal status in the United States, limited English proficiency, and complex ties to labour contractors and foremen, factors that keep many Latino workers from complaining about labour violations or seeking compensation. The vulnerable position of pineros is further exacerbated by the lack of policy and media attention they receive. As noted in Chapter 4, congressional hearings on the labour conditions of agricultural guest workers (H-2B visa holders) in the private sector, while necessary, are particularly ironic given legislators' blindness to the immigrant workforce on *government* land.

Latinos have not been vocal players in debates over forest management in the Rogue Valley. They have little access to the information, organizations, and resources that might help them to better engage with forest management issues. By contrast, white environmental activists and loggers are the most visible constituencies in debates over forest policy. These groups, respectively, have organizations such as Headwaters and the Natural Resources Defense Council, and the Southern Oregon Timber Industries Association and Oregon Association of Loggers to represent their concerns, as well as some access to the political and economic resources (votes and money) needed to make their voices heard.[1]

Some might argue that it is best for undocumented workers *not* to be in the media or policy spotlight so as to avoid deportation and to benefit from access to employment. However, the social and political invisibility of immigrants also allows labour abuse to go unchecked. In Chapter 5, we saw that immigrants have historically improved their working and living conditions only by coming out of the shadows and taking a public stand. The cases of the UFW during the 1960s and PCUN in the 1990s reveal that thousands of farm workers benefited by mobilizing to form a union, bargain collectively, and demand improvements from employers and growers.

Organizing by immigrant workers in urban areas of the United States also points to the potential advantages of vulnerable populations "coming out" in public. For instance, the Justice for Janitors campaign in Los Angeles, spearheaded by the Service Employees International Union during the 1990s, pitted mostly undocumented workers against unscrupulous contractors and building owners. Like pineros, immigrant janitors are employed through contingent contracting and subcontracting arrangements: contractors typically come and go, can fire workers at will, and often change the names of their businesses in order to avoid labour complaints and related legal prosecution. Given the challenges of pinning down wayward contractors, campaign strategists with the Service Employees International Union decided to target owners of the private buildings that janitors were contracted to clean. The logic was that whereas contractors and subcontractors may come and go, building owners cannot so easily walk away from such massive fixed investments.

Being in the public eye was key to the campaign's success. Organizers chose glitzy office buildings in the heart of downtown Los Angeles to make their point. Janitors staged high-profile protests in front of a larger public, which included both numerous white-collar professionals, who were employed by companies that rented space in these buildings, and the media. Building owners were put on the defensive when confronted with the campaign's claims that their multi-million dollar edifices were cleaned in the dark of night by exploited and underpaid immigrant workers. Such public shaming eventually led many building owners to negotiate contracts directly with janitors, thus enabling workers to improve their labour conditions.[2] In this case, the workers' public visibility was crucial to leveraging the situation to their benefit.

In contrast to urban janitors, pineros face unique challenges when it comes to public visibility. As we saw in Chapter 5, the very nature of work in the forest understory – in remote, isolated, and often difficult to access sites – renders pineros largely invisible. Being spread over large areas with little opportunity to communicate with one another poses additional challenges to mobilization efforts. The potential for shaming a "target" as high profile as the federal government, on whose land immigrants labour, is thus compromised by the invisible geography of forest work itself. All else being equal, pineros will likely continue to face labour marginality in forest work.

Policy Implications

The case of pineros indicates the lacunae that still exist and the politics of what is possible on government lands today. Importantly, it also opens a window of insight into gaps in federal policy and in the regulation of labour and economic activity in public forests. The management of federal lands by a low-wage and immigrant labour force has been facilitated through a process of policy fragmentation, whereby unrelated federal programs and policies articulate with decisions of land management agencies to influence the overall management of natural resources on federal lands. Admittedly, some of this policy fragmentation has benefited pineros by allowing them to remain employed regardless of their legal status. Since the late 1980s, for example, Immigration and Naturalization Service authorities have not regularly inspected forest labour crews in Oregon. While this has resulted in fewer worker deportations, other aspects of policy fragmentation have led to continued worker exploitation.

Budget cuts to the Forest Service and Bureau of Land Management, coupled with growing public pressure to fight forest fires in the western United States, have forced agencies to contract out most forest management activities in order to get more done with fewer resources. As this and other studies have shown, contracting arrangements have also exempted government agencies from taking responsibility for labour violations that occur on federal lands.[3] The contracting system ultimately rewards low bidders, often at the expense of lower wages for workers. Labour violations also go unchecked because of budget cuts to state health and labour departments, and a lack of communication between agencies, both of which limit the ability of inspectors to investigate workplace health, safety, and labour conditions.

Contracting may not even be good from a forest health perspective, let alone an economic justice or civil rights viewpoint. When workers face increasingly demanding expectations of faster performance and contractors bid in a highly competitive and cut-throat context, both groups can end up cutting corners. For forest health, this may result in more saplings getting stashed or planted improperly, thus increasing mortality; too many trees of the wrong diameter being thinned in a hurry; and the potentially dangerous situation of losing control over fire when burning piles of brush.

Finally, federal immigration policies such as the 1986 Immigration Reform and Control Act and government programs like the Small Business Administration's set-asides for minority-owned businesses have led to some unintended consequences. The granting of amnesty to a limited pool of applicants allowed for segmentation within manually intensive forestry, with a documented pool of Latino contractors hiring largely undocumented Latino forest workers. Thus, pineros' lack of legal status and close kinship ties to contractors hinders workers from seeking recourse to labour violations.

What then, can be done to effectively improve the working conditions of immigrant forest workers? With immigration reform alternately "on and off" the national agenda, some have suggested amnesty or a path to earned legal status for undocumented workers as a desirable solution. This is in the context of an estimated 12 million undocumented immigrants already living in the United States (many of whom are relatives of naturalized US citizens) and broad public opinion against the mass deportation of all such individuals. Even though some politicians continue to demand extreme policy measures against undocumented groups, mass deportation is ultimately a political and economic non-starter. Not only are naturalized immigrants and their US-born children a growing political constituency for both Republicans and Democrats, but the economic fallout from such a policy would mean that manually intensive forest work and most work in US agriculture and the service sector (secondary labour market) would simply not get done.

In addition, amnesty or earned legal status would mitigate many pineros' fear of deportation. Having legal status may make it easier for some workers to approach government authorities and report unscrupulous contractors, although employees' kinship ties to employers would still pose a barrier to reporting all labour-related violations. Moreover, as we saw in Chapter 3, amnesty without consistent enforcement of employer sanctions could lead to a reprise of the 1990s, which saw the rise of a segmented workforce of documented labour contractors and continued exploitation of undocumented workers.

To effectively protect workers, any future amnesty or earned legalization policy must be accompanied by tough enforcement of both immigration laws and labour laws. Immigration laws must aim to reduce future waves of exploitable immigrants in forest work, with sanctions meted out to employers who hire undocumented workers. However, a strictly one-sided focus on immigration law enforcement is not the entire solution. High-profile workplace raids by the Immigration and Customs Enforcement under the G.W. Bush administration led to the deportation of hundreds of immigrant workers, but it also drove terrified thousands underground. For instance, a raid of the Agriprocessors meat-packing plant in Postville, Iowa, in May 2008 – the largest criminal enforcement operation carried out to date by immigration authorities – detained nearly 400 workers, of whom over 260 pled guilty to using false documents, were imprisoned, and were subsequently deported. The hurried legal proceedings against these workers have been widely criticized by criminal and immigration lawyers for failing to uphold the immigrants' right to due process and were lambasted by social justice groups for tearing apart the social fabric of a rural town.[4]

The Obama administration's approach to undocumented workers has been less severe but continues to reinforce a one-sided focus on immigration law enforcement. The September 2009 firing of eighteen hundred immigrant employees at clothing-maker American Apparel's Los Angeles factory showcased a new strategy for tackling illegal immigration. Rather than using workplace raids, Immigration and Customs Enforcement officials are now forcing employers to dismiss unauthorized workers. "The firings, however, divided opinion in California over the effects of the new approach, especially at a time of high unemployment in the state and with a major, well-regarded employer as a target."[5] Critics also insist that the federal government should focus on penalizing employers who exploit their workers.

Enforcement of labour laws would have to translate into better wages and safer working conditions for all pineros. Such a policy framework, which emphasizes both enforcement and access to legal status, would shift power from contractors – deprived of access to cheap, undocumented labour – to workers protected by labour laws and legal status. With higher wages, safer working conditions, and freedom from the fear of deportation, many pineros would remain forest workers rather than moving on to become labour contractors. With the effective enforcement of immigration policies to limit the entry of unauthorized and vulnerable workers into the US labour market, and the enforcement of labour laws to ameliorate substandard wages and working conditions, there would be little incentive for all legalized workers to become labour contractors.

Government agencies also need to exercise greater oversight over working conditions and immediately penalize labour contractors who violate labour laws. Although federal agencies claim that they are not liable for the labour violations of contractors, the government must take active responsibility for the health and safety of those who help maintain public land. The Forest Service and Department of Labor are currently discussing ways in which they may better co-operate to protect workers' well-being. Again, this would have to mean improved wages, fewer low-bid contracts, and enforced labour protection for all workers, regardless of their legal status. The solution, then, is not just about comprehensive immigration reform. It is also about *comprehensive worker protection.*

By no means are these outcomes inevitable, even with the prospect of legislation on immigration and labour protection. Rather, they are contingent on greater mainstream awareness of forest management and on a demand by the public that forest workers receive just compensation and safe labour conditions. This, in turn, depends on the ability of pineros to forge alliances with environmental and community forestry activists – groups that have successfully pushed their agendas at the state and national levels. Certainly, this is not an impossible endeavour for rural immigrant workers. The successes of the United Farm Workers and Pineros y Campesinos Unidos del Noroeste were largely due to their strategic alliances with student, labour, and religious groups. In the Rogue Valley, immigrant youth are slowly emerging to take on leadership roles in their communities. They have mobilized around access to education and health care and have demanded greater public representation of their histories. It is still too early to say whether these nascent developments will translate into greater recognition for pineros' role in forest management, but given that many of the active youth in the Rogue Valley are also the sons and daughters of forest workers, it is likely that they will give voice to their parents' experiences through their own activism.

Ultimately, the story of forests and forest labour in the US West is about environmental and social justice. Traditionally, environmental justice scholars have focused on the inequitable distribution of environmental harms/toxins in urban areas and the movements led by low-income and people-of-colour communities to overcome these injustices.[6] This book, by contrast, has shown that a vision of environmental well-being, understood in terms of "forest health," is often based on immigrant labour exploitation. Put another way, undocumented Latino men and women physically embody the inequitable distribution of social harm as they maintain public environmental goods.

It is difficult to imagine that manually intensive forest management could be done without the tremendous and invaluable contributions of immigrants. Latinos, and people of colour more generally, are not marginal players in sustaining the health of our environment. By putting their bodies on the line, they are integral to it. Until we publicly acknowledge and drastically improve the poor labour and social conditions of pineros, we must live with the unsustainable truth that national forests in the United States do not represent the "greatest good for the greatest number," but are instead maintained on the backs of abused immigrant workers.

Appendix
Researching *Pineros*

This book is based on field work in Medford, Oregon, and additional archival research at the Bancroft library at the University of California, Berkeley, the Forest History Society in Durham, North Carolina, and the Oregon Historical Society in Portland, Oregon. I used a combination of qualitative research methods including semi-structured interviews, participant observation, analysis of government documents, and content analysis of newspaper coverage.

In total, I conducted over seventy-five interviews with Anglo and Latino forest workers, contractors, and their family members; employees of the US Forest Service, Bureau of Land Management, and Bureau of Labor and Industries; and leaders of local social service and community organizations. I also used snowball sampling to interview a limited number of former tree planters in the Rogue Valley, all of whom were white US citizens. In addition, I obtained interviews with one of the four prominent logging firms in the area, as well as several gyppo loggers who did contract work for this firm. Although these interviews do not represent a saturated sample of the entire population of loggers (company or gyppo) or former reforestation workers, they helped clarify the historical records to which I had access.

Reflecting the composition of the manually intensive forest labour force today, the majority of forest workers and contractors interviewed were Latino males. My interviews with Latino forest workers and their families were conducted mostly in Spanish. I conducted about half of these interviews on my own and the rest with my research assistant, Victoria, who is the daughter of a pinero. Interviews with individuals other than Latino forest workers and their families were conducted in English.

To identify forest labour contractors in the Rogue Valley, I used contractor lists published by the Bureau of Labor and Industries. In addition to in-depth interviews, I gathered information on forest contractors from state and federal records, including the Department of Labor, the Bureau of Labor and Industries contractor lists, the Central Contractor Registry, and the Small Business Administration database. To identify and interview forest workers, I solicited contacts from various people, including government employees, forest labour contractors, and the children of pineros. I also used snowball sampling, whereby forest workers would refer me to other members of their crew.

Both men's and women's narratives were important in constructing a social history of the Latinization of forest work in southern Oregon. Male interviewees generally focused on the details of forest work, while women talked more broadly about living in the Rogue Valley, their interactions with other members of the community, and the challenges their families faced in being among the first Latinos to settle in the area. Some women also touched on the impact of seasonal forest work on families and the role that female relatives played in subsidizing the labour exploitation of men.

I identified social service organizations by asking interviewees about the organizations with which they were familiar and by consulting the Hispanic Yellow Pages. I identified Forest Service and Bureau of Land Management employees through a process of referral by key informants within both of these agencies. Living in Medford also afforded me numerous opportunities for more informal conversations about changes to forest management in the area and about the role of Latinos in both forest work and the larger community. To protect the identities of my respondents, I have used pseudonyms when referring to interviewees and specific contracting firms.

The semi-structured interviews focused on people's experiences in forest work, their relationships to the land, perceptions of and motives for participation in resource management, and their position within social and occupational structures. I conducted semi-structured interviews with the following types of respondents:

1   *Government agency officials (23 interviews).* These include employees of the Bureau of Land Management, the Forest Service, and the Bureau of Labor and Industries. At the Forest Service and Bureau of Land Management, I interviewed employees in various positions, including historians, silviculturists, nursery managers, soil scientists, line

workers, fire fighters, and contracting officers. At the Bureau of Labor and Industries, I interviewed state labour inspectors. I asked questions about changes over the last few decades in land management on public lands and management structures, challenges that agencies face in implementing land management, the composition of forest labour crews, and the nature of working conditions.

2   *Forest labour contractors (10 interviews).* These are individuals licensed by the Bureau of Labor and Industries to operate crews for farm and forest work. To identify and contact forest contractors in the Rogue Valley, I began with contractor lists provided by Oregon's Bureau of Labor and Industries. From a total of nineteen contractors in the Rogue Valley in 2004, I was granted interviews with seven Latino contractors and one Anglo contractor. The two largest forest contractors in the area (in terms of size of their workforces) were both Anglo, but despite numerous attempts, both refused interviews. My questions included the following: How did you enter your line of business? How do you recruit workers? How do you view competition in the industry, and what opportunities for collaboration do you see with federal land management agencies?

3   *Forest workers (28 interviews).* These are individuals who work in the woods as part of a crew or as individuals. I interviewed both foremen, who supervised crews of workers, and workers themselves. My interviews with pineros focused on their experiences in forest management, their motives for working in the woods, their position within social and occupational structures, and their families' histories of settlement in the Rogue Valley. Reflecting the composition of the labour force, the majority of forest workers and contractors interviewed were Latino males. I also interviewed several tree planters who had worked in the woods during the mid-1970s and early 1980s, as well as current-day loggers. Except for two white women who were former tree planters, these interviewees were all white men. My questions included the following: What is the nature of your work and of safety conditions on the job? What challenges do you face in getting access to work, compensation, and benefits? What is the occupational structure? Do you see opportunities for advancement? What is the impact of work on family members?

4   *Family members (21 interviews).* These are spouses and adult children of forest workers and contractors who may do occasional work in the forest. I also interviewed female immigrants, including family members, women who had worked in the woods during the early 1970s, and the first Latina forest contractor in the Rogue Valley. My

questions included the following: What is the social and economic impact of forest work on families? What other areas do family members work in? How do family support structures facilitate the ability of forest workers and contractors to access work? What are the sources of leadership in your community? Who do you go to if you have any problems?

5   *Social service providers and community organizations (22 interviews).* These are employees of local organizations that serve the Latino community. My questions included the following: What services do you provide, and to whom? What are the major concerns you see in the populations you serve? What is your organizational history? What is the present composition of staff and volunteers?

Of the various interviews, those with Latino forest workers and contractors were the most challenging. Many did not understand why I was interested in their stories, nor had they encountered researchers in the past. Forest labour contractors were especially tight-lipped when asked about their work, due to both keen competition between contractors and the undocumented status of many of their employees. Pineros, too, were initially reluctant to talk about their experiences in the woods. Many were related to labour contractors or foremen, and they expressed gratitude simply to have a job.

At the same time, those who did speak at length were extremely generous about sharing their stories and were often encouraged to do so by other members of their families. In this regard, it was the relatives and especially the children of pineros and some labour contractors who were most willing to discuss their families' histories and experiences of work and settlement in the Rogue Valley. For this, I relied on the youths I befriended through my participation in Club Latino, a student-run organization at the local community college in Medford. Many of them were all too familiar with what their relatives endured in the woods and felt it was important to share their stories.

There is little extant written documentation on early Latino settlers in the Rogue Valley, even in institutions designed to house local historical records, such as the Southern Oregon Historical Society. Apart from a few articles on migrant workers in the *Medford Mail Tribune* (the region's major paper), the Southern Oregon Historical Society archives contain little information on the Latino families who first moved into the valley during the 1970s and 1980s. Thus, I constructed much of the history of Latino settlement in the area through interviews with early settlers and members of social service organizations serving the Latino population.

Similarly, no public data was available on the changing ethnic and racial composition of the forest workforce. To document shifts from Anglo to Latino forest workers, I relied on interviews with contractors and workers, complemented by the observations and recollections of Forest Service and Bureau of Land Management employees. In most cases, I was unable to observe crews directly while they worked for three reasons: work sites were located in areas that were extremely isolated and inaccessible by car; labour contractors and foremen were, understandably, unwilling to slow down their work by having a researcher in tow; and Bureau of Land Management and Forest Service officials rarely had the time to visit project sites themselves. However, on the several occasions when I accompanied agency employees to project locations, I was able to observe the ethnic homogeneity and labour conditions of all-Latino crews at work – thinning, piling, and burning brush. These direct observations reinforced the findings from my interview data, all of which point to a homogeneous workforce composed primarily of Mexican immigrants. I also used visits to work sites to recruit pineros and foremen for interviews during their off-work hours.

In Chapter 4, I use three types of indicators to measure the marginality of forest workers: analysis of media coverage, organizational visibility, and engagement in the policy process. In order to gauge the extent to which forest workers are covered in the media, I used electronic archives to examine the number of newspaper articles (without considering length and placement) that discuss loggers, tree planters, and pineros in the Pacific Northwest. At the national level, I examined coverage in the *New York Times* (*NYT*) between 1975 and 2006, and at the local level, I used the *Medford Mail Tribune* (*MMT*), southern Oregon's largest circulating paper. Because the *MMT* electronic archives begin in 1997 and I was denied access to their paper morgues (the paper has not been micro-filmed), my search is limited to *MMT* coverage between 1997 and 2005. I conducted document searches on Lexis/Nexis using terms relevant to each occupational workforce and geographically limited to Oregon, Washington, and the Pacific Northwest. My search for relevant articles in the *NYT* and *MMT* included the following keywords: logging, logger, reforestation, tree-planting, tree-planter, ecosystem management, ecosystem restoration, thinning, immigrant, and pinero. For the *NYT*, I also combined the terms "Latinos and forest," and "Hispanic and forest" in order to narrow down my search parameters. By contrast, the keywords "Latino" and "Hispanic" turned up few articles in the *MMT*, so there was no need to narrow my parameters by combining terms.

Organizations that represent or work with forest workers served as another indicator of the visibility of loggers, tree planters, and pineros to policy makers. I identified such organizations through their coverage in local newspapers, through interviews with forest workers in the Rogue Valley, and by speaking with members of non-profit organizations with which I was already familiar. I also attended public protests over forest management in the Rogue Valley to observe organizational participation or a lack thereof. In the Rogue Valley, forest management issues have largely been framed in terms of whether to log and have mainly involved groups of timber industry advocates, loggers, and environmental activists, all of whom are white.

Because tree planting by Anglo workers waned during the mid-1980s, most of my data on the political visibility of this workforce in Oregon is based on written records from the late 1970s and early 1980s. My main sources include Hal Hartzell's account, *Birth of a Cooperative: Hoedads, Inc.* (1987), and issues of the *ARC Quarterly*, published by the Associated Reforestation Contractors between 1980 and 1985. Written by a former member of the Hoedads tree-planting co-operative and by Anglo reforestation contractors, respectively, these sources document the political involvement of tree-planting co-operatives and reforestation contractors at the level of the state legislature. In addition to relying on these limited historical references, I also interviewed several former Hoedads living in the Rogue Valley, as well as two former reforestation contractors who were founding members of and active participants in the Associated Reforestation Contractors during the late 1970s and early 1980s.

Finally, I relied on academic articles and books to reference some of the ways in which different groups of forest workers and contractors have engaged in the policy process to defend their interests. For example, in the early 1980s, members of tree-planting co-operatives and reforestation contractors clashed over workers' compensation and minimum wage requirements. The mid-1990s saw heated battles between timber groups and environmentalists over habitat conservation for endangered species. With the implementation of the Healthy Forests Restoration Act, conflicts between environmentalists and industry have continued to dominate debates over forest policy in the Pacific Northwest, although some attention has been drawn recently to labour conditions among forest workers.

The overall dearth of documentation about pineros reflects a larger lack of information about the manually intensive aspects of forest management. Whether it is hand-planting cutover areas, thinning overcrowded stands of small-diameter trees, or pruning dense underbrush, our national forests are actively maintained by *people*, most of whom we never see. Only a few written sources document the existence of earlier groups of Anglo reforestation workers,[1] and the invisibility of forest labour continues today. With the exception of a significant literature on the reforestation efforts of the Civilian Conservation Corps during the Great Depression, the story of non-logging manual labour in the woods has been given relatively short shrift. While the occasional news article or public hearing may pull the cover back on the exploitation of forest workers, the status quo for pineros remains one of labour invisibility and political quiescence.

Notes

Preface

1 See the Appendix for details of the methods used in this study.
2 Studs Terkel, *Hard Times* (New York: Pantheon Books, 1970), 3.

Chapter 1: Invisible Workers

1 Ken Kesey, *Sometimes a Great Notion: A Novel* (New York: Viking Press, 1964).
2 Robert E. Swanson, *Rhymes of a Western Logger* (Vancouver, BC: Lumberman Printing, 1943), 6.
3 William G. Robbins, *Hard Times in Paradise: Coos Bay, Oregon, 1850-1986* (Seattle: University of Washington Press, 1988).
4 Matthew S. Carroll et al., "Adaptation Strategies of Displaced Idaho Woods Workers: Results from a Longitudinal Panel Study," *Society and Natural Resources* 13 (2000): 95-113; Steven E. Daniels, Corinne L. Gobeli, and Angela J. Findley, "Reemployment Programs for Dislocated Timber Workers: Lessons from Oregon," *Society and Natural Resources* 13 (2000): 135-50; J. Kusel et al., "Effects of Displacement and Outsourcing on Woods Workers and Their Families," *Society and Natural Resources* 13 (2000): 115-34.
5 To protect the identity of individuals, I use pseudonyms throughout this book.
6 Richard Hansis, "The Harvesting of Special Forest Products by Latinos and Southeast Asians in the Pacific Northwest: Preliminary Observations," *Society and Natural Resources* 9 (1996): 611-15; Rebecca J. McLain and Eric Jones, *Challenging "Community" Definitions in Sustainable Natural Resource Management: The Case of Wild Mushroom Harvesting in the USA* (London: International Institute for Environment and Development, 1997).

    Salal harvesters include Latino immigrant floral-greens (or "brush") harvesters on the Olympic Peninsula of Washington, where floral-greens, salal, or "brush" harvesting is done on both public and private lands. These greens are collected by individuals who purchase harvesting permits from private landholders, from land management agencies such as the Forest Service, or from "brush sheds" who have negotiated harvesting rights from landowners. Harvesters then sell their brush to "buyers" or "brush sheds," who go on to sell the greens in the international market. At present, floral-greens harvesters are considered independent contractors rather than employees. For more on floral-greens harvesting, see

Heidi Ballard and Lynn Huntsinger, "Salal Harvester Local Ecological Knowledge, Harvest Practices and Understory Management on the Olympic Peninsula, Washington," *Human Ecology* 34 (2006): 529-47; Heidi Ballard, "Impacts of Harvesting Salal (Gaultheria shallon) on the Olympic Peninsula, Washington: Harvester Knowledge, Science and Participation" (PhD diss., University of California, Berkeley, 2004); Sarah Loose, "The Workers behind the Wreaths," *Jefferson Center News* 4, 2 (2005): 1-3.

7 Daniel Kemmis, *Community and the Politics of Place* (Norman: University of Oklahoma Press, 1992); James McCarthy, "Neoliberalim and the Politics of Alternatives: Community Forestry in British Columbia and the United States," *Annals of Association of American Geographers* 96, 1 (2006): 84-104.

8 Thomas Brendler and Henry Carey, "Community Forestry, Defined," *Journal of Forestry* 96, 3 (1998): 21-23; Richard Pardo, "Community Forestry Comes of Age," *Journal of Forestry* 93, 11 (1995): 20-24.

9 Gerald J. Gray, "Understanding Community-Based Forest Ecosystem Management: An Editorial Synthesis," *Journal of Sustainable Forestry* 12, 3 (2001): 1-23.

10 Jeffrey G. Borchers and Jonathan Kusel, "Toward a Civic Science for Community Forestry," in *Community Forestry in the United States: Learning from the Past, Crafting the Future,* ed. Mark Baker and Jonathan Kusel (Washington, DC: Island Press, 2003), 147-63; Edward Weber, "A New Vanguard for the Environment: Grass-Roots Ecosystem Management as a New Environmental Movement," *Society and Natural Resources* 13 (2000): 237-59.

11 Teresa Satterfield, *Anatomy of a Conflict* (Vancouver: UBC Press, 2002); J. LeMonds, *Deadfall: Generations of Logging in the Pacific Northwest* (Missoula, MT: Mountain Press Publishing, 2001); Matthew Carroll, *Community and the Northwest Logger* (Boulder, CO: Westview Press, 1995); Robbins, *Hard Times in Paradise.*

12 William Cronon, "The Trouble with Wilderness; or Getting Back to the Wrong Nature," in *Uncommon Ground: Rethinking the Human Place in Nature* (New York: W.W. Norton, 1996), 69-90.

13 This evolutionary development, as elaborated by Turner,

> begins with the Indian and the hunter; it goes on with the disintegration of savagery by the entrance of the trader ... the pastoral stage in ranch life; the exploitation of the soil by the raising of unrotated crops of corn and wheat in sparsely settled farm communities; the intensive culture of the denser farm settlement; and finally the manufacturing organization with the city and the factory system. Frederick Jackson Turner, Chapter 1 in *The Frontier in American History* (New York: Henry Holt, 1921), http://xroads.virginia.edu/~HYPER/TURNER/, last modified 30 September 1997.

14 See Patricia Nelson Limerick, *The Legacy of Conquest: The Unbroken Past of the American West* (New York: Norton, 1987); William Cronon, George Miles, and Jay Gitlin, "Becoming West: Toward a New Meaning for Western History," in *Under an Open Sky: Rethinking America's Western Past,* ed. William Cronon, George Miles, and Jay Gitlin (New York: W.W. Norton, 1992), 3-27; Amy Kaplan, "Manifest Domesticity," *American Literature* 70, 3 (1998): 581-606.

15 Tomás Almaguer, *Racial Fault Lines: The Historical Origins of White Supremacy in California* (Berkeley: University of California Press, 1994), 210.

16 See, for example, M. Hill, "Adding Diversity to the Outdoors," *NY Newsday*, 25 August 2005; Sonya Geis, "Forest Service Faulted for Lack of Outreach Programs," *Pasadena Star-News*, 5 December 2004; Carolyn Finney, "Black Faces, White Spaces: African Americans and the Great Outdoors," *Community Forestry Newsletter* (Winter 2004): 2-4.

17 Jeff Romm, "The Coincidental Order of Environmental Justice," in *Justice and Natural Resources: Concepts, Strategies and Applications*, ed. Kathryn M. Mutz, Gary C. Bryner, and Douglas S. Kenney (Washington, DC: Island Press, 2002), 117-38.

18 Carolyn Merchant, "Shades of Darkness: Race and Environmental History," *Environmental History* 8 (2003): 380-94; Geoff Mann, "Race, Skill, and Section in Northern California," *Politics and Society* 30, 3 (2002): 465-96; Geoff Mann, "The State, Race, and 'Wage Slavery' in the Forest Sector of the Pacific North-West United States," *Journal of Peasant Studies* 29, 1 (2001): 61-88; Beverly Brown and Agueda Marin-Hernandez, *Voices from the Woods: Lives and Experiences of Non-Timber Forest Workers* (Portland, OR: Jefferson Center for Education and Research, 2001).

19 Jorge Durand, Douglas S. Massey, and Fernando Charvet, "The Changing Geography of Mexican Immigration to the United States: 1910-1996," *Social Science Quarterly* 81, 1 (2000): 1-15.

20 William Kandel and Emilio Parrado, "Hispanics in the American South and the Transformation of the Poultry Industry," in *Hispanic Spaces, Latino Places*, ed. Daniel Arreola (Austin: University of Texas Press, 2004), 255-76; Ann V. Millard and Jorge Chapa, eds., *Apple Pie and Enchiladas: Latino Newcomers in the Rural Midwest* (Austin: University of Texas Press, 2004); L. Stephen, *The Story of PCUN and the Farmworker Movement in Oregon,* in collaboration with PCUN staff members (Eugene: University of Oregon, Department of Anthropology, 2001); Erasmo Gamboa and Carolyn Buan, *Nosotros: The Hispanic People of Oregon* (Portland: Oregon Council for Humanities, 1995).

21 Alejandro Portes and Rubén G. Rumbaut, *Immigrant America: A Portrait*, 3rd ed. (Berkeley: University of California Press, 2006); Audrey Singer, Susan Wiley Hardwick, and Caroline Brettell, *Twenty-First-Century Gateways: Immigrant Incorporation in Suburban America* (Washington, DC: Brookings Institution Press, 2008).

22 Tom Knudson and Hector Amezcua, "The Pineros: Men of the Pines," *Sacramento Bee*, November 2005.

23 Brinda Sarathy and Vanessa Casanova, "Guest Workers or Unauthorized Immigrants? The Case of Forest Workers in the United States," *Policy Sciences* 41, 2 (2008): 95-114.

24 Kirk Johnson, "With Illegal Immigrants Fighting Wildfires, West Faces a Dilemma," *New York Times*, 28 May 2006; Craig Welch, "A War in the Woods," *Seattle Times*, 6 June 2006.

25 Josh McDaniel and Vanessa Casanova, "Forest Management and the H2B Guest Workers Program in the Southeastern United States: An Assessment of Contractors and Their Crews," *Journal of Forestry* 103, 3 (2005): 114-19; Jim Hamilton, "Feliz Navidad? Labor and Perspective in North Carolina's Christmas Tree Industry," paper presented at the Annual Meeting of the Rural Sociological Society, Montreal,

Canada, 27-29 June 2003; John Bliss, Tamara Walkingstick, and Conner Bailey, "Development or Dependency? Sustaining Alabama's Forest Communities," *Journal of Forestry* 96, 3 (1998): 24-30.

26 Ballard and Huntsinger, "Salal Harvester Local Ecological Knowledge"; McLain, *Challenging "Community" Definitions*; Hansis, "Harvesting of Special Forest Products."

27 McDaniel and Casanova, "Forest Management and the H-2B Guest Workers Program."

28 Hamilton, "Feliz Navidad?" The H-2B visa guest worker program was implemented after the Immigration Reform and Control Act of 1986. It is intended to prohibit employers from hiring illegal, undocumented workers. The H-2B visa is for non-professional, non-agricultural workers for jobs lasting less than one year. These H-2B "non-immigrant" workers are employed in a variety of industries ranging from forestry and poultry in the Southeast to service positions in hotels in coastal and resort cities. H-2A visas, while similar to H-2B visas, are used for agricultural labour, which includes Christmas tree plantations. The difference in classification between Christmas tree work and tree planting is related to Internal Revenue codes and US Department of Agriculture categories. A US Department of Labor directive notes that the difference between the two occupations is bureaucratic rather than necessarily related to differences in the actual type of work: "Although the occupations of Tree Planter and Laborer, Brush Clearing have many similarities to agriculture, they are not so classified under either the Internal Revenue Code or the Fair Labor Standards Act." Distinctions between forest and field work are thus reinforced and reproduced even within the bureaucracy of visa categories. See US Department of Labor, "Procedures for H-2B Temporary Labor Certification in Nonagricultural Occupations," General Administration Letter No. 01-95, 22 December 1997, http://www.ows.doleta.gov/dmstree/gal/gal95/gal_01-95c1.htm.

29 Brendan Sweeney, "Sixty Years on the Margin: The Evolution of Ontario's Tree Planting Industry and Labour Force, 1945-2007," *Labour/Le Travail* 63 (2009): 47-78.

30 Mann, "State, Race, and 'Wage Slavery'"; Brinda Sarathy, "The Latinization of Forest Management Work in Southern Oregon: A Case from the Rogue Valley," *Journal of Forestry* 104, 7 (2006): 359-65; Sarathy and Casanova, "Guest Workers or Unauthorized Immigrants?"

31 Cassandra Moseley, *Procurement Contracting in the Affected Counties of the Northwest Forest Plan: Twelve Years of Change* (Portland, OR: US Department of Agriculture, Forest Service, PNW Research Station, 2006).

32 Jackson County GIS Services, telephone interview, 2 February 2006.

33 Mann, "State, Race, and 'Wage Slavery.'"

34 Alejandro Portes and Robert L. Bach, *Latin Journey: Cuban and Mexican Immigrants in the United States* (Berkeley: University of California Press, 1985); Portes and Rumbaut, *Immigrant America*.

35 Cecilia Menjívar, *Fragmented Ties: Salvadoran Immigrant Networks in America* (Berkeley: University of California Press, 2000); Cecilia Menjívar, "Impact of the Receiving Context: Salvadorans in San Francisco in the Early 1990s," *Social Problems* 44 (1997): 104-23.

36 Vanessa Casanova and Josh McDaniel, "No Sobra y No Falta: Recruitment Networks and Guest Workers in Southeastern US Forest Industries," *Urban Anthropology and Studies of Cultural Systems and World Economic Development* 34 (2005): 45-84.

37 Michael Piore, *Birds of Passage: Migrant Labor and Industrial Societies* (Cambridge, UK: Cambridge University Press, 1979); Connor Bailey, "Segmented Labor Markets in Alabama's Pulp and Paper Industry," *Rural Sociology* 61, 3 (1996): 475-96.

38 Sweeney draws upon Jamie Peck's causal emphases of segmentation theory. See Jamie Peck, *Work-Place: The Social Regulation of Labor Markets* (New York: Guilford Press, 1996); Sweeney, "Sixty Years on the Margin."

Chapter 2: Cutting and Planting

1 US Census Bureau, "The 2010 Statistical Abstract: Population," last modified 27 October 2010, http://www.census.gov/compendia/statab/2010/cats/population.html.

2 Michael Williams, *Americans and Their Forests: A Historical Geography* (Cambridge, UK: Cambridge University Press, 1989), 160.

3 Richard Rajala, "The Forest as Factory: Technological Change and Worker Control in the West Coast Logging Industry, 1880-1930," *Labour/Le Travail* 32 (1993): 73-104; William G. Robbins, *Lumberjacks and Legislators: Political Economy of the US Lumber Industry, 1890-1941.* (College Station: Texas A&M University Press, 1982).

4 William Greeley, *Forests and Men* (Garden City, NY: Doubleday, 1951).

5 Ibid., 41.

6 "Land Fraud in the West: Extensive 'Graft' Schemes Discovered on the Pacific Coast," *New York Times*, 22 October 1903.

7 Williams, *Americans and Their Forests*, 161.

8 Williams, *Americans and Their Forests*; Robbins, *Lumberjacks and Legislators*.

9 Richard W. Massey, "A History of the Lumber Industry in Alabama and West Florida, 1880-1914" (PhD diss., Vanderbilt University, 1960).

10 Williams, *Americans and Their Forests*, 238.

11 Ibid., 264.

12 Ibid., 324.

13 W. Scott Prudham, *Knock on Wood: Nature as Commodity in Douglas-Fir Country* (New York: Routledge, 2005).

14 Emily Brock, "The Challenge of Reforestation: Ecological Experiments in the Douglas Fir Forest, 1920-1940," *Environmental History* 9, 1 (2004): 1-21.

15 An open stand is one where the forest structure is primarily composed of well-spaced and large-diameter, mature trees. Historically, ponderosa pine forests were open-stand systems, partially maintained through Native American practices of burning underbrush on the forest floor. William G. Robbins and Donald Wolf, *Landscape and the Intermontane Northwest: An Environmental History* (Portland, OR: Pacific Northwest Research Station, General Technical Report PNW-GTR-319, 1994).

16 Richard Rajala, *Clearcutting the Pacific Rain Forest: Production, Science, and Regulation.* (Vancouver: UBC Press, 1998), 92.

17 Ibid., 100.

18   Ibid., 93.
19   Ibid., 101.
20   Samuel T. Dana and Sally Fairfax, *Forest and Range Policy*, McGraw-Hill Series in Forest Resources (New York: McGraw-Hill, 1980), 52.
21   Nancy Langston, "Forest Dreams, Forest Nightmares: An Environmental History of a Forest Health Crisis," in *American Forests: Nature, Culture, Politics*, ed. Char Miller (Lawrence: University of Kansas Press, 1997), 250.
22   For more on the selective logging vs. clear-cutting debates during this period, see Rajala, *Clearcutting the Pacific Rain Forest*, 125-32.
23   Carlile Winslow, "Wood and War," *Journal of Forestry* 40, 12 (1942): 920-22.
24   Arthur Upson, "Our Forest Resources Are Contributing to Victory," *Journal of Forestry* 40, 12 (1942): 913.
25   William Robbins, "Lumber Production and Community Stability: A View from the Pacific Northwest," *Journal of Forest History* 31 (1987): 194; David Correia, "The Sustained Yield Forest Management Act and the Roots of Environmental Conflict in Northern New Mexico," *Geoforum* 38 (2007): 1040-51.
26   Robbins, "Lumber Production and Community Stability," 196.
27   Paul Hirt, *A Conspiracy of Optimism: Management of National Forests since World War Two* (Lincoln: University of Nebraska Press, 1994), 45.
28   William G. Robbins, *Landscapes of Conflict: The Oregon Story, 1940-2000*, Weyerhaeuser Environmental Classics (Seattle: University of Washington Press, 2004), 168.
29   By 1970, the cut from federal land was 18 percent of the total timber harvest. Robbins, *Lumberjacks and Legislators*, 247.
30   Robbins, *Landscapes of Conflict*, 161.
31   Ibid., 186.
32   Interview with Jeffrey Lelande, Rogue River-Siskiyou Forest Service historian, 29 May 2004, Medford, OR.
33   Robbins, *Landscapes of Conflict*, 176.
34   Greeley, *Forests and Men*, 134.
35   Rajala, *Clearcutting the Pacific Rain Forest*, 134-35.
36   David Wilma, "Weyerhaeuser Dedicates the Nation's First Tree Farm near Montesano on June 21, 1941," HistoryLink.org, Essay 5256, 21 February 2003, http://www.historylink.org.
37   Richard E. McArdle and Walter H. Meyer, *The Yield of Douglas-Fir in the Pacific Northwest*, Forest Service Technical Bulletin (Washington, DC: US Department of Agriculture, Forest Service, 1930); Brian Cleary, Robert Greaves, and Richard Hermann, "Regenerating Oregon's Forests: A Guide for the Regeneration Forester" (Corvallis: Oregon State University, 1978); Brock, "Challenge of Reforestation."
38   Rajala, *Clearcutting the Pacific Rain Forest*, 144. Some years after Olzendam coined the phrase "tree farm," Arthur Priaulx, a public-relations man for the West Coast Lumberman's Association, called on the public to turn the Tillamook Burn (a series of forest fires between 1933 and 1951 that burned 355,000 acres of old-growth forest in Oregon's Tillamook State Forest) into "a vast 300,000-acre tree farm." Priaulx also came up with the idea of taking Oregon school children to the woods to help replant the burn. Between 1950 and 1970, more than 25,000 school children participated in this effort under the auspices of the "Plant Trees

and Grow Citizens" program. Gail Wells, *The Tillamook: A Created Forest Comes of Age* (Corvallis: Oregon State University Press, 1999), 58-59.

39 "History of the American Tree Farm System," *Western Conservation Journal* 23, 2 (1966): 47.

40 For more details on industry's public relations initiatives and the consequent public enthusiasm for the Tree Farm Program, see Rajala, *Clearcutting the Pacific Rain Forest*, 173-75.

41 Cleary, Greaves, and Hermann, "Regenerating Oregon's Forests," 3.

42 Bob Zybach, "Renewed Resources: The Reforestation of the Tillamook Burn," *ARC Quarterly* (Fall 1983): 13-17.

43 Hirt, *A Conspiracy of Optimism*, 204-5.

44 Ibid., 118.

45 Ibid., 207.

46 Ibid., 212.

47 Greeley, *Forests and Men*, 167-68.

48 Otis, Alison T. et al., *The Forest Service and the Civilian Conservation Corps: 1933-42* (Washington, DC: US Department of Agriculture, Forest Service, August 1986), http://www.nps.gov/history/history/online_books/ccc/ccc/index.htm.

49 Wilmon H. Droze, *Trees, Prairies, and People: A History of Tree Planting in the Plain States* (Denton, TX: Texas Woman's University, 1977).

50 "ARC Focus: Bob Snow," *ARC Quarterly* (Spring 1984): 21-22.

51 Telephone interview with Ryan Thomas (pseudonym), 18 May 2004, Medford, OR.

52 "ARC Focus: Bob Snow." Pulling green chain at a mill and setting chokers on landing sites were considered the lowest-paying jobs in the timber industry.

53 Telephone interview with Ryan Thomas (pseudonym), 18 May 2004, Medford, OR. Bob Snow's real name is used since it appears within *ARC Quarterly* publications.

54 "ARC Focus: Bob Snow." Stashing refers to the practice of ditching trees. Since workers in the industry were initially paid by the piece, it was often profitable for workers and contractors to increase their tally by throwing away, rather than planting, trees.

55 Within the United States, fourteen states have forest practices acts but only six have comprehensive ones that support state enforcement, education of operators, monitoring, and complete administration of the act. Others call for using "best management practices" on a voluntary or required basis without the enforcement and education component. In other states, a variety of regulatory mechanisms and agencies are involved in regulating forest practices. John Garland, "The Oregon Forest Practice Act: 1972-1994," in Forest Codes of Practice: Contributing to Environmentally Sound Forest Operations, FAO Forestry Paper, 1996, FAO Corporate Document Repository, http://www.fao.org/docrep/w3646e/w3646e07.htm.

56 T. Lorensen, "The Forest Practices Act: Protecting Resources for 30 Years," *Forest Log* (July-August 2001): 14. The 1941 Oregon Forest Conservation Act required timber operators to withhold a small amount of timber for natural reforestation of cutover areas, or, if the area was to be clear-cut, deposit a bond in the state treasury to finance reforestation by artificial means. The legislation did not ensure

the successful development of a second commercial crop and did not comply with scientific research on the requirements necessary for abundant reforestation. Shaped by industry interests, the Oregon Forest Conservation Act was more a symbol of forest conservation in the fight against federal intervention. For more on the inability of state legislation to regulate industry's cutting practices, see Rajala, *Clearcutting the Pacific Rain Forest*, 169-89.

57 Hal Hartzell, *Birth of a Co-operative: Hoedads, Inc.* (Eugene: Hologos'i, 1987), 27-28.

58 For more on the links between technological change and maintaining control over workers in forestry, see Rajala, "Forest as Factory."

59 Planters carry a hoedad (or hoedag), a curved, round-nosed blade used to dig holes twelve to fifteen inches deep. The curved blade was introduced around 1976. Prior to that time, the hoedad had a flat blade. Thank you to Dean Pihlstrom for this clarification. Slash is woody debris, such as tree bark, broken tree limbs, and other organic material left over from logging.

60 Interview with Raul Fernández (pseudonym), ex-foreman and tree planter, 25 March 2004, Medford, OR.

61 Hartzell, *Birth of a Co-operative*, 35.

62 Hartzell, *Birth of a Co-operative*, 307.

63 Ibid., 41.

64 *Reforestation Efforts in Western Oregon: Hearing before the Subcommittee on Forests of the Committee on Agriculture, House of Representatives, Ninety-Fifth Congress, First Session, July 8, 1977, Roseburg, Oreg.* (Washington D.C., Government Printing Office, 1977).

65 Co-operative members were not the only ones who considered tree planting "more than a job." One former reforestation contractor aptly noted:

> A very large percentage of contractor crew workers also very much considered their employment more than a job, though for generally different reasons. Many private crews were essentially families – though if you pressed me on my language, I would say they had characteristics more like gangs, though without violent acts. A lot of generally alienated and almost unemployable young men found (and here I am exaggerating a bit for effect) fathers in their foremen. The foremen controlled almost every aspect of their lives. Living in camps deep in the forest, with the foremen controlling the only vehicles, these young men worked hard, drank hard, ate very well, and slept 9 hours per night in their warm tents. I have seen countless 18-19 year old lost souls, pack on 30 pounds of hard muscle, learn unbending work habits, save money, and emerge in their mid-twenty's [sic] as self confident and very employable men with a lifetime of memories packed into five years time. Email, 17 September 2010.

66 Hartzell, *Birth of a Co-operative*, 21.

67 Ibid., 323.

68 Ibid., 324-25.

69 Gerald Mackie, "Success and Failure in an American Workers' Co-operative Movement," *Politics and Society* 22, 2 (1994): 215-35.

70 Hartzell, *Birth of a Cooperative*, 337.

71 Mackie, "Success and Failure," 94.

72 The Hoedads inspired a number of other tree-planting co-operatives in the Pacific Northwest during the course of the 1970s and early 1980s, including Second Growth, Green Side Up, Thumb Reforestation, and Marmot. For more information about the rise and fall of tree-planting co-operatives, see Mackie, "Success and Failure."

73 Mackie, "Success and Failure," 226.

74 J. Stauffer, "The President's Side," *ARC Quarterly* (Fall/Winter 1984): 4.

75 Telephone interview with one of the charter members of ARC, 4 September 2005. I was fortunate to be able to track down this individual twenty years after the demise of ARC.

76 For more information on the details of ARC's lawsuit against the State of Oregon for failing to make co-operatives comply with labour laws equally, see "Changes in Tax Law," *ARC Quarterly* (Fall 1981): 9-10. Reprinted from National Federation of Independent Business Research and Education Foundation.

77 Steven Winston, "A Letter to the Editor," *ARC Quarterly* (Fall 1983): 7.

78 "What is A.R.C.?" *ARC Quarterly* (Spring 1982): 26.

79 Jack Watson, "Minimum Wage," *ARC Quarterly* (Spring/Summer 1985): 5.

80 Problems with workers' compensation in Oregon extended well beyond the reforestation industry. During the 1980s, Oregon businesses had some of the highest workers' compensation costs in the United States, while Oregon workers had among the highest rates of injury and illness nation-wide. In 1990, Ted Kulongoski (current Oregon governor), then serving as the state's insurance commissioner, spearheaded an overhaul of the workers' compensation system. Commonly referred to as the "Mahonia Hall Reforms," the new legislation both lowered workers' compensation rates and increased benefits for workers. http://www.orosha. org/admin/newsrelease/2010/nr2010_13.pdf.

81 Bob Zybach, "Safety and Profit," *ARC Quarterly* (Fall 1981): 12.

82 Ibid.

83 Stauffer, "President's Side," 4.

84 Stauffer, "President's Side," 6.

85 *ARC Quarterly* (Spring 1982): 18.

86 *ARC Quarterly* (Spring 1982): 5.

87 Bob Erny was the US Forest Service official in charge of the contract, and Sharipoff was a forest labour contractor in Oregon during the early 1980s. The phrase "Erny defaulted Sharipoff" meant that Erny terminated Sharipoff's contract with the Forest Service. Scott Coleman, "Two Busts = Default!! Big Deal!" *ARC Quarterly* (Spring 1982): 13.

88 I discuss the Bracero Program in more detail in the next chapter.

89 Mackie, "Success and Failure," 229; Prudham, *Knock on Wood*, 49.

Chapter 3: From Pears to Pines

1 "Best Places to Retire: Medford, Oregon," *CNNMoney.com*, 5 May 2003, http:// money.cnn.com/2002/05/01/retirement/bpretire_medford/; "Best Places to Retire 2005: Where You'll Live," *CNNMoney.com*, http://money.cnn.com/magazines/ moneymag/bpretire/2005/index.html.

2 Kay Atwood, *Blossoms and Branches: A Gathering of Rogue Valley Orchard Memories* (Ashland, OR: Gandee Printing Center, 1980), 1.

3   Don Mitchell, *The Lie of the Land: Migrant Workers and the California Landscape* (Minneapolis: University of Minneapolis Press, 1996); Carey McWilliams, *Factories in the Field: The Story of Migratory Farm Labor in California* (Santa Barbara: Peregrine Publishers, 1971).

4   Micah Bump, Lindsey B. Lowell, and Silje Pettersen, "The Growth and Population Characteristics of Immigrants and Minorities in America's New Settlement States," in *Beyond the Gateway: Immigrants in a Changing America*, ed. Elzbieta M. Godziak and Susan Forbes Martin (Lanham, MD: Lexington Books, 2005), 19-53.

5   J. Cromartie, "Race and Ethnicity in Rural Areas," *Rural Conditions and Trends* 9, 2 (1999): 9-19.

6   Kandel and Parrado, "Hispanics in the American South"; Durand, Massey, and Charvet, "Changing Geography of Mexican Immigration."

7   The Rogue Valley includes the following towns and cities in Jackson County, Oregon: Ashland, Central Point, Eagle Point, Medford, Phoenix, Talent, and White City.

8   Robert Bernstein, "Hispanic Population Passes 40 Million, Census Bureau Reports," 9 June 2005, http://www.census.gov/newsroom/releases/archives/population/cb05-77.html.

9   Arthur D. Murphy, Colleen B. Blanchard, and Jennifer A. Hill, eds., *Latino Workers in the Contemporary South* (Athens: University of Georgia Press, 2001); Millard and Chapa, *Apple Pie and Enchiladas*.

10 David Griffith, "Hay Trabajo: Poultry Processing, Rural Industrialization and the Latinization of Low-Wage Labor," in *Any Way You Cut It: Meat Processing and Small Town America*, ed. Donald D. Stull, Michael J. Broadway, and David Griffith (Lawrence: University of Kansas Press, 1995), 129-51; Kandel and Parrado, "Hispanics in the American South"; Dell Champlin and Eric Hake, "Immigration as Industrial Strategy in American Meatpacking," *Review of Political Economy* 18, 1 (2006): 49-69; Donald D. Stull, Michael J. Broadway, and Ken C. Erickson, "The Price of a Good Steak: Beef Packing and Its Consequences for Garden City, Kansas," in *Structuring Diversity: Ethnographic Perspectives on the New Immigration*, ed. Louise Lamphere (Chicago: University of Chicago Press, 1992), 35-64; Rubén Hernández-Léon and Victor Zúñiga, "'Making Carpet by the Mile': The Emergence of a Mexican Immigrant Community in an Industrial Region of the US Historic South," *Social Science Quarterly* 81, 1 (2000): 49-65.

11 Durand, Massey, and Charvet, "Changing Geography of Mexican Immigration."

12 Stephen, *Story of PCUN;* Gamboa and Buan, *Nosotros.*

13 *Federal Real Property Profile* (Washington, DC: General Service Administration Office of Government Policy, 30 September 2004), www.census.gov/compendia/statab/2007/tables/07s0346.xls.

14 Mario Compean, "Mexican Americans in the Columbia Basin: Historical Overview," Washington State University Library Archive, accessed 10 July 2006, http://www.vancouver.wsu.edu/crbeha/ma/ma.htm.

15 Francisco E. Balderrama and Raymond Rodriguez, *Decade of Betrayal: Mexican Repatriation in the 1930s* (Albuquerque, NM: University of New Mexico Press, 1995).

16 Erasmo Gamboa, *Mexican Labor and World War II: Braceros in the Pacific Northwest, 1942-47* (Seattle/Austin: University of Washington Press/University of Texas Press, 1990), 49.

17 "Migrant Camp Is Solicited Here," *Medford Mail Tribune,* 18 July 1941.
18 "Mexicans Will Work Here until First of Year," *Medford Mail Tribune,* 9 November 1945.
19 Gamboa, *Mexican Labor and World War II,* 19.
20 "Labor Camps Rise in Northwest as Opposition Wanes," *Medford Mail Tribune,* 7 July 1939.
21 Fred L. Koestler, "Bracero Program," *Handbook of Texas Online* (http://tshaonline. org/handbook/online/articles/omb01), accessed 3 April 2006. Published by the Texas State Historical Association.
22 Juan Ramon García, *Operation Wetback: The Mass Deportation of Mexican Undocumented Workers in 1954* (Westport, CT: Greenwood Press, 1980).
23 Douglas S. Massey, *Beyond Smoke and Mirrors: Mexican Immigration in an Era of Economic Integration* (New York: Russell Sage Foundation, 2002).
24 Susan Ferriss and Ricardo Sandoval, *The Fight in the Fields: Cesar Chavez and the Farmworkers Movement,* ed. Diana Hembree (San Diego, CA: Harcourt Brace, 1997).
25 David Rothenberg, *With These Hands: The Hidden World of Migrant Farmworkers Today* (Berkeley: University of California Press, 2000).
26 For more on the conditions in labour camps in the Rogue Valley, see "Harvesting Hopes," *Medford Mail Tribune,* 6 September 1981.
27 Oral history account of Ned Vilas, orchard owner, in Atwood, *Blossoms and Branches,* 185.
28 Marjorie Edens, "Oral History Interview with A.C. Allen, Jr., and His Wife Eve," Jacksonville Museum, Oregon, 26 November 1979, 48-49, Southern Oregon Historical Society, Tape 125.
29 Interview, 26 February 2004, Medford, OR.
30 Edens, "Oral History Interview," 48.
31 Scholars have debated whether Latino immigrant arrivals are simply relocating from other parts of the United States or arriving in non-gateway states like Oregon directly from abroad. While the latter may be the case for immigrant arrivals in the 1990s and later, earlier waves of Latinos to non-gateway states may have relocated from other parts of the United States. Certainly, this was the case for María López and her family. Her account reveals that immigrants of her father's generation often worked in California and other states before eventually moving to Oregon and settling in the Rogue Valley. See Elzbieta M. Gozdziak and Susan Forbes Martin, *Beyond the Gateway: Immigrants in a Changing America* (Lanham, MD: Lexington Books, 2005); William Frey, "US Census Shows Different Paths for Domestic and Foreign-Born Migrants," *Population Today* 30, 6 (2002): 1, 4-5; Bump, Lowell, and Pettersen, "Growth and Population Characteristics," 33.
32 Interview with María López (pseudonym), 26 April 2004, Medford, OR.
33 Interview with Pedro Sánchez (pseudonym), 1 July 2004, Central Point, OR.
34 Interview with Alejandra Puentes (pseudonym), 20 April 2004, Medford, OR.
35 Interview with Lucía Tomás (pseudonym), 20 July 2004, Medford, OR.
36 Interview with Lydia Rincón (pseudonym), 28 April 2004, Medford, OR.
37 Interview with Lupe Ramos (pseudonym), 27 September 2004, Medford, OR.
38 Interview with Ester Moraga (pseudonym), 22 July 2004, Medford, OR.
39 Interview, 13 April 2004, Medford, OR.
40 Interview with Gustavo Flores (pseudonym), 2 August 2004, Medford, OR.
41 Interview with Braulio Santos (pseudonym), 21 September 2004, Medford, OR.

42 "Jim Daniels" and "Rogue Valley Reforestation" are both pseudonyms.
Records for 1980 were obtained from Oregon's Bureau of Labor and Industries.
Record keeping at the time lumped together both farm labour and tree-planting
contractors, so it was not possible to determine those who hired only farm
workers, those who hired workers exclusively in forestry, and those who hired
both types of workers. Early records also include tree-planting co-operatives
under the category of labour contractors, even though co-operatives did not
technically have employee-employer structures. Given these record-keeping
practices, the actual number of forest labour contractors in Oregon was probably
well below 106.

43 To "scalp" a tree is to clear the ground of any vegetation at the base of where a
sapling is to be planted. This limits competition from weeds and other vegetation
and increases a sapling's chance of survival.

44 Interview with Rubén Gómez (pseudonym), 25 March 2004, Medford, OR.

45 The number of Latino forest contractors in the Rogue Valley is derived from
public records of Oregon's Bureau of Labor and Industries using counts of Hispanic
surnames with further refinement through informant interviews. Given the lack
of public data on undocumented immigrants, the share of such workers in the
forest labour force is based on estimates provided by forest contractors and Forest
Service employees, which have ranged from 90 to 95 percent.

46 Ecosystem Workforce Program, "Data Provided by Ecosystem Workforce Program
on 1998 and 1999 Total Market Share of Region 6 Forest Service Awarded Forest
Management Contracts" (University of Oregon, 2004).

47 I began with a list of all the forest management contracts awarded by the Region
6 Forest Service in 1998 and 1999 provided by the Ecosystem Workforce Program.
To create a list of Oregon-based forest contractors, I cross-referenced all the names
on the Ecosystem Workforce Program list with names of licensed forest contractors
on Oregon Bureau of Labor and Industries' lists. Names appearing on both lists
then constituted the list of Oregon-based contractors.

48 Interview with Alberto Morales (pseudonym), 17 June 2004, Medford, OR.

49 Interview with María Morales (pseudonym), 17 June 2004, Medford, OR.

50 Champlin and Hake, "Immigration as Industrial Strategy in American Meat-
packing," 65.

51 Interview, 20 April 2004, Medford, OR.

52 Interview, 7 August 2004, Medford, OR.

53 The 8(a) contractors in the Rogue Valley have done especially well. For instance,
three of the top five contracts awarded in 1998-99 went to 8(a) Latino contractors
from the Rogue Valley. Ecosystem Workforce Program, "Data Provided by Ecosystem Workforce Program."

54 Interview with Maggie Giuliani, Rogue River-Siskiyou Forest Service contracting
officer, 7 July 2004, Medford, OR.

55 The role of labour exploitation through family and social networks is addressed
further in Chapter 4.

56 Max J. Pfeffer, "Low-Wage Employment and Ghetto Poverty: A Comparison of
African-American and Cambodian Day-Haul Farm Workers in Philadelphia,"
Social Problems 41, 1 (1994): 9-29.

57 Rachel Carson's *Silent Spring* (1962), on the deleterious effects of DDT on plant
and animal life, was one of the most influential pieces to inspire a growing environmental movement in the United States.

58 The most significant pieces of legislation from the 1970s include the National Environmental Policy Act (1970), the Endangered Species Act (1973), the National Forest Management Act (1976), and the Federal Land Policy and Management Act (1976). Dana and Fairfax, *Forest and Range Policy.*

59 William C. Boyd, "New South, New Nature: Regional Industrialization and Environmental Change in the Post-New Deal American South" (PhD diss., University of California, Berkeley, 2002).

60 Oregon Department of Consumer and Business Services, *Occupational Safety and Health in Oregon's Forests: Logging and Forestry Services* (Salem, OR: Oregon Department of Consumer and Business Services, December 2005); Cassandra Moseley, "Who Works in the Woods?" (paper presented at the "The Northwest Forest Plan: Ten Years Later," Portland, OR, 13 April 2004).

61 James Beltram, Rick Evans, Michael Hibbard, and James Luzzi. "The Scope and Future Prospects: Oregon's Ecosystem Management Industry," EWP Working Paper No. 1 (Ecosystem Workforce Program, Institute for a Sustainable Environment, University of Oregon, Fall 2001).

62 Richard W. Haynes and Gloria E. Perez, *Northwest Forest Plan Research Synthesis* (Portland, OR: US Department of Agriculture, Forest Service, PNW Research Station, 2000).

63 Oregon Department of Forestry, "Figure 38-2. Estimated Annual Reforestation Spending 1982 Dollars," accessed 16 May 2011, http://www.oregon.gov/ODF/STATE_FORESTS/FRP/crt6ind38.shtml.

64 In 2002, the Biscuit fire burned over 500,000 acres of forest in southern Oregon and northern California. Arizona, Colorado, and Oregon also recorded their largest fires of the last hundred years in 2002. According to end-of-year agency reports, wildland fires burned a total of 7,186,368 acres across the United States in 2002. http://www.nifc.gov/fire_info/ytd_state_2002.htm.

65 The Healthy Forests Restoration Act (HFRA) of 2003 has been extremely controversial. Aimed at aggressively suppressing forest fires, the act seeks to expedite the ability of federal land management agencies to thin overstocked forest stands, remove excess vegetation, and eliminate fuel loads. To accomplish these objectives, the HFRA legislation revised regulations implemented by the National Forest Management Act, the National Environmental Policy Act, and the Endangered Species Act, key pieces of environmental legislation. Environmental groups in particular argue that the HFRA is designed to produce greater timber harvests from public lands rather than to promote healthy forests. For more on the HFRA controversy, see Jacqueline Vaughn and Hanna J. Corter, *George W. Bush's Healthy Forests: Reframing the Environmental Debate* (Boulder: University Press of Colorado, 2005); Jesse B. Davis, "The Healthy Forests Initiative: Unhealthy Policy Choices in Forest Fire Management," *Environmental Law* 34 (2004): 1209-45.

66 Cassandra Moseley, "Creating Community Benefit," EWP Briefing Paper Number 5 (Institute for a Sustainable Environment, Ecosystem Workforce Program, University of Oregon, 2004).

67 Beltram and Evans, "Scope and Future Prospects."

68 Mark Baker, "Socioeconomic Characteristics of the Natural Resources Restoration System in Humboldt County," 2003, http://www.sierrainstitute.us/HTML/Publications.html#anchor.

69 Oregon Department of Consumer and Business Services, *Occupational Safety and Health in Oregon's Forests.*

70   Cassandra Moseley and Stacey Shankle, "Who Gets the Work? National Forest Contracting in the Pacific Northwest," *Journal of Forestry* 99, 9 (2001): 32-37.

Chapter 4: The Marginality of Forest Workers

1   Interview, 5 August 2004, Medford, OR.
2   Tim Costello, "The Once and Future History of Steady Work," *Working USA* 4, 3 (2000): 86-99; Catherine Ruckelshaus and Bruce Goldstein, *The Legal Landscape for Contingent Workers in the United States* (New York: National Employment Law Project, Farmworker Justice Fund, 2002); A. Kalleberg, B. Reskin, and K. Hudson, "Bad Jobs in America: Standard and Nonstandard Employment Relations and Job Quality in the United States," *American Sociological Review* 65, 2 (2000): 256-78; A. Kalleberg, "Nonstandard Employment Relations: Part-Time, Temporary, and Contract Work," *Annual Review of Sociology* 26 (2000): 341-65.
3   Catherine Ruckelshaus and Bruce Goldstein, "From Orchards to the Internet: Confronting Contingent Worker Abuse," National Employment Law Project, 2002, http://www.nelp.org/content/content_publications/P600/.
4   Beverly A. Brown et al., *Contract Forest Labourers in Canada, the US, and Mexico* (Wolf Creek, OR: Jefferson Center for Education and Research, 2004).
5   Rural Voices Conservation Coalition, "Workforce and Labour Issue Paper," Sustainable Northwest, May 2006, http://www.sustainablenorthwest.org/resources/rvcc-issue-papers.
6   Kalleberg, Reskin, and Hudson, "Bad Jobs in America."
7   Miriam Wells, "Immigration and Unionization in the San Francisco Hotel Industry," in *Organizing Immigrants: The Challenge for Unions in Contemporary California* (Ithaca, NY: ILR Press, 2000), 109-29.
8   Knudson and Amezcua, "Pineros: Men of the Pines."
9   Carroll, *Community and the Northwest Logger.*
10   LeMonds, *Deadfall.*
11   Haynes and Perez, *Northwest Forest Plan.*
12   Satterfield, *Anatomy of a Conflict.*
13   Richard Rajala, "Bill and the Boss: Labour Protest, Technological Change, and the Transformation of the West Coast Logging Camp, 1890-1930," *Journal of Forest History* 33, 4 (1989): 179.
14   Rajala, "Bill and the Boss."
15   Mann, "Race, Skill, and Section."
16   Jerry Lembcke and William Tattam, *One Union in Wood* (Madeira Park, BC/New York: Harbour Publishing/International Publishers, 1984), 18-19.
17   For more details on the 1935 lumber strike, see Lembcke and Tattam, *One Union in Wood*, 30-43; Robert Ficken, *The Forested Land: The History of Lumbering in Western Washington* (Durham/Seattle: Forest History Society/University of Washington Press, 1987), 209-17; Carroll, *Community and the Northwest Logger;* LeMonds, *Deadfall.*
18   LeMonds, *Deadfall*, 2.
19   Kim Voss and Rick Fantasia, *Hard Work: Remaking the American Labor Movement.* (Berkeley: University of California Press, 2004).
20   Robbins, *Lumberjacks and Legislators.*
21   Prudham, *Knock on Wood;* Carroll et al., "Adaptation Strategies"; Kusel et al., "Effects of Displacement and Outsourcing."

22 William Dietrich, *The Final Forest: The Battle for the Last Great Trees of the Pacific Northwest* (New York: Penguin, 1992); Satterfield, *Anatomy of a Conflict*; Robbins, *Lumberjacks and Legislators.*

23 Daniels, Gobeli, and Findley, "Reemployment Programs"; Kusel et al., "Effects of Displacement and Outsourcing"; Carroll et al., "Adaptation Strategies."

24 Bureau of Labor Statistics, *National Census of Fatal Occupational Injuries in 2005* (Washington, DC: US Department of Labor, 2006), http://www.bls.gov/iif/oshcfoi1.htm.

25 Bureau of Labor Statistics, *Workplace Injuries and Illnesses in 2004* (Washington, DC: US Department of Labor, 17 November 2005), http://www.bls.gov/iif/home.htm.

26 Oregon Department of Consumer and Business Services, *Occupational Safety and Health in Oregon's Forests*; Bureau of Labor Statistics, *Workplace Injuries and Illnesses in 2004.*

27 Oregon Department of Consumer and Business Services, *Occupational Safety and Health in Oregon's Forests.*

28 Knudson and Amezcua, "Pineros: Men of the Pines." Many of the fatalities reported on by Knudson and Amezcua involved Latino floral-greens harvesters, who are considered independent contractors. The *Sacramento Bee* reporters did not specify whether any fatalities involved forest workers contracted to work on federal lands.

29 Agricultural Worker Health Project, "Farmworker Transportation," accessed 31 December 2009, http://www.agworkerhealth.org/RTF1.cfm?pagename=Transportation.

30 Bureau of Labor Statistics, *Workplace Injuries and Illnesses in 2004.* This figure is based on national data, which categorized non-fatal injuries according to SOC (Standard Occupational Classification) codes. "Logger" includes the SOC occupational categories of "faller" and "equipment operator." "Tree-planter" and "ecosystem management worker" share the same SOC occupational category of "forest and conservation worker."

31 In Oregon alone, there were over one hundred disabling claims for forestry services workers in 2004. Oregon Department of Consumer and Business Services, *Occupational Safety and Health in Oregon's Forests.*

32 Ibid. "Bodily reaction" refers to injuries such as strains and sprains, sustained as a result of actions like bending, reaching, twisting, or slipping without falling.

33 Ibid.

34 Nancy Corrin, "Spinal Column," *ARC Quarterly* (Summer 1983): 13-14.

35 US Department of Labor, Bureau of Labor Statistics, "May 2005 State Occupational Employment and Wage Estimates," last modified 24 May 2006, http://www.bls.gov/oes/2005/may/oessrcst.htm. There was no wage data for the SOC categories of "fallers" or "forest conservation workers" for Oregon.

36 Bureau of Labor Statistics, "Table 10. Private Industry by State and Six-Digit NAICS Industry: Establishments, Employment, and Wages, 2004 Annual Averages," accessed 3 May 2009, http://www.bls.gov/cew/ew04sector11.pdf.

37 US Department of Labor, *Register of Wage Determinations under the Service Contract Act* (Washington, DC: US Department of Labor, 2004).

38 US Department of Agriculture, "Migrant and Seasonal Agricultural Worker Protection Act," last modified 30 September 2002, http://www.thecre.com/fedlaw/

legal19/mspasumm.htm. For more details, see Jack, L. Runyan, *A Summary of Federal Laws and Regulations Affecting Agricultural Employers, 1992* (Washington, DC: US Department of Agriculture, Agriculture and Rural Economy Division, Economic Research Service, 1992).

39 Cassandra Moseley, "Ethnic Differences in Job Quality among Contract Forest Workers on Six National Forests," *Policy Sciences* 3, 2 (2006): 115.
40 Hartzell, *Birth of a Co-operative*, 33-34.
41 Mackie, "Success and Failure," 226.
42 LeMonds, *Deadfall*, 82-87.
43 Moseley, "Ethnic Differences in Job Quality," 113.
44 William Gamson and Andre Modigliani, "Media Discourse and Public Opinion on Nuclear Power: A Constructionist Approach," *American Journal of Sociology* 95, 1 (1989): 1-37; Myra M. Ferree et al., *Shaping Abortion Discourse* (Cambridge, UK: Cambridge University Press, 2002).
45 Vaughn and Corter, *George W. Bush's Healthy Forests*.
46 Refer to the Appendix for further details on the methods used for this section's media analysis.
47 Sarathy, "Latinization of Forest Management Work."
48 Moseley, *Procurement Contracting*.
49 *Forest Service Workers* (Washington, DC: Government Printing Office, 2006).
50 Johnson, "With Illegal Immigrants Fighting Wildfires"; Welch, "A War in the Woods."
51 Interview with Rob Pointer (pseudonym), 10 August 2004, Medford, OR.
52 Forest Community Research, *Assessment of the Northwest Economic Adjustment Initiative* (Taylorsville, CA: Forest Community Research, December 2002), http://www.sierrainstitute.us/neai/NEAIAssessment.html.
53 LeMonds, *Deadfall*.
54 Interview with Mollie Owens-Stevenson, Jobs in the Woods coordinator, 21 October 2004, Rogue Community College, Medford, OR.
55 Daniels, Gobeli, and Findley, "Reemployment Programs."
56 Hartzell, *Birth of a Co-operative*, 337-38.
57 Interview with Valentín (pseudonym), 10 August 2004, Phoenix, OR.
58 Loose, "Workers behind the Wreaths."
59 More information about the forum can be found at http://ewp.uoregon.edu/resources/workforce.
60 Kate Bagby and Jonathan Kusel, "Civic Science Partnerships in Community Forestry: Building Capacity for Participation among Underserved Communities" (Taylorsville, CA: Pacific West Community Forest Center, Forest Community Research, 2003); Beverly Brown, *Challenges Facing Community Forestry: The Role of Low-Income Forest Workers* (Wolf Creek, OR: Jefferson Center for Education and Research, 2001); Debra J. Salazar and Donald K. Alper, *Sustaining the Forests of the Pacific Coast* (Vancouver: UBC Press, 2001).
61 Stephen, *Story of PCUN*.
62 Under federal law, it is illegal for undocumented immigrants to work in the United States and employers are prohibited from knowingly hiring them. Once hired, however, most states (with the exception of Wyoming) do provide workers' compensation benefits regardless of immigration status. See, for example, Mark Noonan, "Raising Debate beyond Borders," *Risk & Insurance,* 3 March 2011;

National Immigration Law Project, "California and Maryland Courts Uphold Undocumented Immigrants' Rights to Workers' Compensation," *Immigrants' Rights Update* 19, 2 (2005).

63 Differences in legal status between undocumented Latino workers and documented Latino contractors raise interesting questions about the class position of Latino contractors as a whole. While contractors clearly are employers, they are not capitalists in the traditional sense. Contractors hire labourers to provide forestry services on federal land but do not own the means of production (land or equipment). As noted earlier, pineros often provide their own tools, including chainsaws and safety gear. Contractors thus serve as a convenient interface to institutionalize an ironic employment relationship between the state and the undocumented workers it criminalizes.

64 Norman S. Hayner, "Taming the Lumberjack," *American Sociological Review* 10, 2 (1944): 217-25; Ralph W. Andrews, *Glory Days of Logging* (Seattle: Superior Publishing, 1956); LeMonds, *Deadfall*. See also Steve Maynard, "Rough Work and Rugged Men: The Social Construction of Masculinity in Working-Class History," *Labour/Le Travail* 23 (1989): 159-69.

65 Brinda Sarathy, "Minutes from Meeting between Matsutake Mushroom Harvesters and US Forest Service, Crescent Lake, OR," 30 July 2002.

66 Sheila Foster, "Environmental Justice in an Era of Devolved Collaboration," in *Justice and Natural Resources: Concepts, Strategies and Applications*, ed. Kathryn M. Mutz, Gary C. Bryner, and Douglas S. Kenney (Washington, DC: Island Press, 2002), 139-60.

67 Loose, "Workers behind the Wreaths."

68 Catherine A. Lutz and Jane L. Collins, *Reading National Geographic* (Chicago: University of Chicago Press, 1993).

69 Scott T. Smith, "Word of Mouth: Las Morenas," *Medford Mail Tribune*, 20 January 2006.

70 Sarah Lemon, "Nopales Grill," *Medford Mail Tribune*, 3 June 2005.

71 John Darling, "Life and Death: A Celebration," *Medford Mail Tribune*, 2 November 2005.

72 City of Medford Council meeting minutes, 21 April 2005.

73 Historic Commission minutes, 7 June 2005, Medford, OR.

74 Paris Achen, "We Are Here to Work: Students March to Support Illegal Immigrants," *Medford Mail Tribune*, 8 April 2006.

75 John Darling, "We Are Not Criminals," *Medford Mail Tribune*, 1 April 2006.

Chapter 5: A Tale of Two Valleys

1 Rick Tejada-Flores and Ray Telles, *The Fight in the Fields: Cesar Chavez and the Farmworkers' Struggle*, directed by Rick Tejada-Flores and Ray Telles, produced by Paradigm Productions, a presentation of the Independent Television Service, San Francisco, CA, 1997.

2 Ruth Milkman, *Organizing Immigrants: The Challenge for Unions in Contemporary California* (Ithaca, NY: ILR Press, 2000).

3 John D. McCarthy and Mayer Zald, "Resource Mobilization and Social Movements: A Partial Theory," *American Journal of Sociology* 82, 6 (1977): 1212-41. The resource-mobilization model in social movement theory has been rightly critiqued for its overemphasis on elites and has been displaced by a political-process model. The

latter highlights attributes internal to movements, such as indigenous leadership and resources, and their relation to external factors. Doug McAdam, John D. McCarthy, and Mayer Zald, eds., *Comparative Perspectives on Social Movements: Political Opportunities, Mobilizing Structures, and Cultural Framings* (Cambridge, UK: Cambridge University Press, 1996).

4 David Harvey, *A Brief History of Neoliberalism* (Oxford: Oxford University Press, 2005).

5 Michael Piore and Charles F. Sabel, *The Second Industrial Divide: Possibilities for Prosperity* (New York: Basic, 1984).

6 In this scenario, timing carries most of the explanatory weight for variations in organizational outcome. This is not to discount the role of other factors in influencing outcomes. Variations in leadership style, strategic capacity, involvement of movement participants themselves, and "mobilizing culture" (ideas and practices around appropriate mobilizing goals and strategies) are also important considerations. Scholar Heidi Swarts has shown that both organizational culture (which emphasizes the internal processes and characteristics of an institution) *and* the political-economic context (timing may be seen as part of this larger context) enable or constrain particular outcomes. Heidi Swarts, *Organizing Urban America* (Minneapolis: University of Minnesota Press, 2008). However, given limited access to information about the internal culture of defunct Rogue Valley organizations profiled in this chapter, I focus largely on the role of timing and the broader political-economic context in determining organizational outcomes.

7 Ferriss and Sandoval, *Fight in the Fields.*

8 J. Craig Jenkins and Charles Perrow, "Insurgency of the Powerless: Farm Worker Movements (1946-1972)," *American Sociological Review* 42, 2 (1977): 253.

9 Ibid., 264.

10 Ibid., 262.

11 Doug McAdam, *Political Process and the Growth of Black Insurgency, 1930-1970* (Chicago: University of Chicago Press, 1982); Marshall Ganz, "Resources and Resourcefulness: Strategic Capacity in the Unionization of California Agriculture, 1959-1966," *American Journal of Sociology* 105, 4 (2000): 1003-62.

12 Ganz, "Resources and Resourcefulness," 1005.

13 Laura Pulido, *Environmentalism and Economic Justice: Two Chicano Struggles in the Southwest* (Tucson: University of Arizona Press, 1996).

14 Jenkins and Perrow, "Insurgency of the Powerless," 263.

15 In 2002, the UFW board shifted from its historic focus on organizing agricultural workers to the Latino population in general. The union has been accused of enriching members of the Chavez family while no longer helping the majority of farm workers. In 2006, the UFW did not have a single contract in the table grape vineyards of its birthplace in the Central Valley. Miriam Pawel, "UFW: A Broken Contract," *Los Angeles Times*, 8 January 2006.

16 Gamboa and Buan, *Nosotros.*

17 Stephen, *Story of PCUN*, 11.

18 L. Stephen, "Cultural Citizenship and Labor Rights for Oregon Farmworkers: The Case of Pineros y Campesinos Unidos del Noroeste (PCUN)," *Human Organization* 62, 1 (2003): 27-38.

19 Stephen, *Story of PCUN.*

20 Ibid.; Stephen, "Cultural Citizenship and Labor Rights."
21 Telephone interview with Tim Buckley (pseudonym), 9 April 2008.
22 "Farmworkers Strike Threat," *Medford Mail Tribune*, 1 July 1981.
23 Ibid.
24 Edens, "Oral History Interview."
25 Telephone interview with Tim Buckley (pseudonym), 9 April 2008.
26 Gamboa and Buan, *Nosotros*, 51-52.
27 Interview, 26 April 2004, Medford, OR.
28 Interview, 13 April 2004, Medford, OR.
29 Interview, 12 April 2004, Medford, OR.
30 Interview, 28 April 2004, Medford, OR.
31 "La Clinica fact sheet," accessed 10 May 2011, http://www.laclinicahealth.org/News.asp.
32 Interview, 12 April 2004, Medford, OR.
33 Interview, 28 April 2004, Medford, OR.
34 There is one immigrant social justice organization in the Rogue Valley, Unete, based in Medford. Unete works in conjunction with immigrant groups in the Willamette Valley and has primarily focused on national campaigns such as comprehensive immigration reform and the passage of the Dream Act. Most of Unete's funding comes from state and national foundations, and the group, which was labelled "too radical" by many residents, receives limited local support.
35 McAdam, *Political Process*.
36 L. Stephen, *Transborder Lives: Indigenous Oaxacans in Mexico, California, and Oregon* (Durham, NC: Duke University Press, 2007); Stephen, *Story of PCUN*.
37 As noted in Chapter 4, most of the contemporary activism around forest worker issues has been spearheaded by mid-level non-profits. Organizations such as the Jefferson Center for Education and Research, the Alliance of Forest Workers and Harvesters, and the Sierra Institute have brought together scholars, policy makers, land managers, and forest workers and harvesters to engage with issues of concern to workers. While these meetings have been incredibly productive in raising awareness and, in some cases, changing decision-making processes to be more inclusive of workers, these organizations have typically not been able to improve fundamental labour conditions related to wages, safety, and exploitation.
38 S. Karthick Ramakrishnan and Paul George Lewis, *Immigrants and Local Governance: The View from City Hall* (San Francisco: Public Policy Institute Of California, 2005).
39 Kristi Andersen, *New Immigrant Communities: Finding a Place in Local Politics* (Boulder, CO: Lynne Rienner Publishers, 2010); Ramakrishnan and Lewis, *Immigrants and Local Governance*.
40 Andersen, *New Immigrant Communities*.
41 Eric Camayd-Freixas, "Interpreting after the Largest ICE Raid in US History: A Personal Account," 13 June 2008, http://graphics8.nytimes.com/packages/pdf/national/20080711IMMIG.pdf; Austin Jenkins, "Rumors and Panic Follow Immigration Raids," *KUOW Program Archive*, 27 June 2007, http://www.kuow.org/defaultProgram.asp?ID=13001.

Chapter 6: Conclusion

1 Steven Davis, "Environmental Politics and the Changing Context of Interest Group Organization," *Social Science Journal* 33, 4 (1996): 343-57.
2 Milkman, *Organizing Immigrants.*
3 Prudham, *Knock on Wood.*
4 Camayd-Freixas, "Interpreting after the Largest ICE Raid"; Julia Preston, "Immigrants' Speedy Trials after Raid Become Issue," *New York Times*, 8 August 2008; Julia Preston, "Iowa Rally Protests Raid and Conditions at Plant," *New York Times*, 28 July 2008.
5 Julia Preston, "Immigration Crackdown with Firings, Not Raids," *New York Times*, 29 September 2009.
6 Robert Bullard, *Dumping in Dixie: Race, Class, and Environmental Quality* (Boulder, CO: Westview Press, 1994); David Pellow, *Garbage Wars: The Struggle for Environmental Justice in Chicago* (Cambridge, MA: MIT Press, 2002).

Appendix: Researching Pineros

1 Here too, I had to rely on a combination of interviews with former contractors, ex-planters, and government employees. I was fortunate to have been able to interview one of the first Anglo forest labour contractors in southern Oregon before he passed away.

Bibliography

Achen, Paris. "We Are Here to Work: Students March to Support Illegal Immigrants." *Medford Mail Tribune*, 8 April 2006.

Agricultural Worker Health Project. "Farmworker Transportation." Accessed 31 December 2009. http://www.agworkerhealth.org/RTF1.cfm?pagename= Transportation.

Almaguer, Tomás. *Racial Fault Lines: The Historical Origins of White Supremacy in California*. Berkeley: University of California Press, 1994.

Andersen, Kristi. *New Immigrant Communities: Finding a Place in Local Politics*. Boulder, CO: Lynne Rienner Publishers, 2010.

Andrews, Ralph W. *Glory Days of Logging*. Seattle: Superior Publishing, 1956.

"ARC Focus: Bob Snow." *ARC Quarterly* (Spring 1984): 21-22.

Atwood, Kay. *Blossoms and Branches: A Gathering of Rogue Valley Orchard Memories*. Ashland, OR: Gandee Printing Center, 1980.

Bagby, Kate, and Jonathan Kusel. "Civic Science Partnerships in Community Forestry: Building Capacity for Participation among Underserved Communities." Taylorsville, CA: Pacific West Community Forest Center, Forest Community Research, 2003.

Bailey, Connor. "Segmented Labor Markets in Alabama's Pulp and Paper Industry." *Rural Sociology* 61, 3 (1996): 475-96.

Baker, Mark. "Socioeconomic Characteristics of the Natural Resources Restoration System in Humboldt County." 2003. http://www.sierrainstitute.us/HTML/ Publications.html#anchor.

Balderrama, Francisco E., and Raymond Rodriguez. *Decade of Betrayal: Mexican Repatriation in the 1930s*. Albuquerque: University of New Mexico Press, 1995.

Ballard, Heidi. "Impacts of Harvesting Salal (*Gaultheria shallon*) on the Olympic Peninsula, Washington: Harvester Knowledge, Science and Participation." PhD diss., University of California, Berkeley, 2004.

Ballard, Heidi, and Lynn Huntsinger. "Salal Harvester Local Ecological Knowledge, Harvest Practices and Understory Management on the Olympic Peninsula, Washington." *Human Ecology* 34 (2006): 529-47.

Beltram, James, Rick Evans, Michael Hibbard, and James Luzzi. "The Scope and Future Prospects: Oregon's Ecosystem Management Industry." EWP Working

Paper No. 1. Ecosystem Workforce Program, Institute for a Sustainable Environment, University of Oregon, Fall 2001.

Bernstein, Robert. "Hispanic Population Passes 40 Million, Census Bureau Reports." 9 June 2005. http://www.census.gov/newsroom/releases/archives/population/cb05-77.html.

"Best Places to Retire: Medford, Oregon." *CNNMoney.com*. 5 May 2003. http://money.cnn.com/2002/05/01/retirement/bpretire_medford/.

"Best Places to Retire 2005: Where You'll Live." *CNNMoney.com*. http://money.cnn.com/magazines/moneymag/bpretire/2005/index.html.

Bliss, John, Tamara Walkingstick, and Conner Bailey. "Development or Dependency? Sustaining Alabama's Forest Communities." *Journal of Forestry* 96, 3 (1998): 24-30.

Borchers, Jeffrey G., and Jonathan Kusel. "Toward a Civic Science for Community Forestry." In *Community Forestry in the United States: Learning from the Past, Crafting the Future*, edited by Mark Baker and Jonathan Kusel, 147-63. Washington, DC: Island Press, 2003.

Boyd, William C. "New South, New Nature: Regional Industrialization and Environmental Change in the Post-New Deal American South." PhD diss., University of California, Berkeley, 2002.

Brendler, Thomas, and Henry Carey. "Community Forestry, Defined." *Journal of Forestry* 96, 3 (1998): 21-23.

Brock, Emily. "The Challenge of Reforestation: Ecological Experiments in the Douglas Fir Forest, 1920-1940." *Environmental History* 9, 1 (2004): 1-21.

Brown, Beverly. *Challenges Facing Community Forestry: The Role of Low-Income Forest Workers*. Wolf Creek, OR: Jefferson Center for Education and Research, 2001.

Brown, Beverly, and Agueda Marin-Hernandez. *Voices from the Woods: Lives and Experiences of Non-Timber Forest Workers*. Portland, OR: Jefferson Center for Education and Research, 2000.

Brown, Beverly A., Diana Leal-Mariño, Kirsten McIllveen, Ananda L. Tan, and Sarah Loose. *Contract Forest Laborers in Canada, the US, and Mexico*. Wolf Creek, OR: Jefferson Center for Education and Research, 2004.

Bullard, Robert. *Dumping in Dixie: Race, Class, and Environmental Quality*. Boulder, CO: Westview Press, 1994.

Bump, Micah, Lindsey B. Lowell, and Silje Pettersen. "The Growth and Population Characteristics of Immigrants and Minorities in America's New Settlement States." In *Beyond the Gateway: Immigrants in a Changing America*, edited by Elzbieta M. Godziak and Susan Forbes Martin, 19-53. Lanham, MD: Lexington Books, 2005.

Bureau of Labor Statistics. *National Census of Fatal Occupational Injuries in 2005*. Washington, DC: US Department of Labor, 2006. http://www.bls.gov/iif/oshcfoi1.htm.

–. "Table 10. Private Industry by State and Six-Digit NAICS Industry: Establishments, Employment, and Wages, 2004 Annual Averages." Accessed 3 May 2003. http://www.bls.gov/cew/ew04sector11.pdf.

–. *Workplace Injuries and Illnesses in 2004*. Washington, DC: US Department of Labor, 17 November 2005. http://www.bls.gov/iif/home.htm.

Camayd-Freixas, Eric. "Interpreting after the Largest ICE Raid in US History: A Personal Account." 13 June 2008. http://graphics8.nytimes.com/packages/pdf/national/20080711IMMIG.pdf.

Carroll, Matthew. *Community and the Northwest Logger.* Boulder, CO: Westview Press, 1995.

Carroll, Matthew S., Keith A. Blatner, Frederick Alt, Ervin G. Schuster, and Angela J. Findley. "Adaptation Strategies of Displaced Idaho Woods Workers: Results from a Longitudinal Panel Study." *Society and Natural Resources* 13 (2000): 95-113.

Casanova, Vanessa, and Josh McDaniel. "No Sobra y No Falta: Recruitment Networks and Guest Workers in Southeastern US Forest Industries." *Urban Anthropology and Studies of Cultural Systems and World Economic Development* 34 (2005): 45-84.

Champlin, Dell, and Eric Hake. "Immigration as Industrial Strategy in American Meatpacking." *Review of Political Economy* 18, 1 (2006): 49-69.

"Changes in Tax Law." *ARC Quarterly* (Fall 1981): 9-10. Reprinted from National Federation of Independent Business Research and Education Foundation.

Cleary, Brian, Robert Greaves, and Richard Hermann. "Regenerating Oregon's Forests: A Guide for the Regeneration Forester." Corvallis: Oregon State University, School of Forestry, 1978.

Coleman, Scott. "Two Busts = Default!! Big Deal!" *ARC Quarterly* (Spring 1982): 13.

Compean, Mario. "Mexican Americans in the Columbia Basin: Historical Overview." Washington State University Library Archive. Accessed 10 July 2006. http://www.vancouver.wsu.edu/crbeha/ma/ma.htm.

Correia, David. "The Sustained Yield Forest Management Act and the Roots of Environmental Conflict in Northern New Mexico." *Geoforum* 38 (2007): 1040-51.

Corrin, Nancy. "Spinal Column." *ARC Quarterly* (Summer 1983): 13-14.

Costello, Tim. "The Once and Future History of Steady Work." *Working USA* 4, 3 (2000): 86-99.

Cromartie, J. "Race and Ethnicity in Rural Areas." *Rural Conditions and Trends* 9, 2 (1999): 9-19.

Cronon, William. "The Trouble with Wilderness; or Getting Back to the Wrong Nature." In *Uncommon Ground: Rethinking the Human Place in Nature*, 69-90. New York: W.W. Norton, 1996.

Cronon, William, George Miles, and Jay Gitlin. "Becoming West: Toward a New Meaning for Western History." In *Under an Open Sky: Rethinking America's Western Past*, edited by William Cronon, George Miles, and Jay Gitlin, 3-27. New York: W.W. Norton, 1992.

Dana, Samuel T., and Sally Fairfax. *Forest and Range Policy.* McGraw-Hill Series in Forest Resources. New York: McGraw-Hill, 1980.

Daniels, Steven E., Corinne L. Gobeli, and Angela J. Findley. "Reemployment Programs for Dislocated Timber Workers: Lessons from Oregon." *Society and Natural Resources* 13 (2000): 135-50.

Darling, John. "Life and Death: A Celebration." *Medford Mail Tribune*, 2 November 2005.

–. "We Are Not Criminals." *Medford Mail Tribune*, 1 April 2006.

Davis, Jesse B. "The Healthy Forests Initiative: Unhealthy Policy Choices in Forest Fire Management." *Environmental Law* 34 (2004): 1209-45.

Davis, Steven. "Environmental Politics and the Changing Context of Interest Group Organization." *Social Science Journal* 33, 4 (1996): 343-57.

Dietrich, William. *The Final Forest: The Battle for the Last Great Trees of the Pacific Northwest.* New York: Penguin, 1992.

Droze, Wilmon H. *Trees, Prairies, and People: A History of Tree Planting in the Plain States*. Denton, TX: Texas Woman's University, 1977.

Durand, Jorge, Douglas S. Massey, and Fernando Charvet. "The Changing Geography of Mexican Immigration to the United States: 1910-1996." *Social Science Quarterly* 81, 1 (2000): 1-15.

Ecosystem Workforce Program. "Data Provided by Ecosystem Workforce Program on 1998 and 1999 Total Market Share of Region 6 Forest Service Awarded Forest Management Contracts." University of Oregon, 2004.

Edens, Marjorie. "Oral History Interview with A.C. Allen, Jr., and His Wife Eve." Jacksonville Museum, OR, 26 November 1979. Southern Oregon Historical Society, Tape 125.

Federal Real Property Profile. Washington, DC: General Service Administration Office of Government Policy, 30 September 2004. www.census.gov/compendia/statab/2007/tables/07s0346.xls.

Ferree, Myra M., William Gamson, Jurgen Gerhards, and Dieter Rucht. *Shaping Abortion Discourse*. Cambridge, UK: Cambridge University Press, 2002.

Ferriss, Susan, and Ricardo Sandoval. *The Fight in the Fields: Cesar Chavez and the Farmworkers Movement*. Edited by Diana Hembree. San Diego: Harcourt Brace, 1997.

Ficken, Robert. *The Forested Land: The History of Lumbering in Western Washington*. Durham/Seattle: Forest History Society/University of Washington Press, 1987.

Finney, Carolyn. "Black Faces, White Spaces: African Americans and the Great Outdoors." *Community Forestry Newsletter* (Winter 2004): 2-4.

Forest Community Research. *Assessment of the Northwest Economic Adjustment Initiative*. Taylorsville, CA: Forest Community Research, 2002.

Forest Service Workers. Washington, DC: Government Printing Office, 2006.

Foster, Sheila. "Environmental Justice in an Era of Devolved Collaboration." In *Justice and Natural Resources: Concepts, Strategies and Applications*, edited by Kathryn M. Mutz, Gary C. Bryner, and Douglas S. Kenney, 139-60. Washington, DC: Island Press, 2002.

Frey, William. "US Census Shows Different Paths for Domestic and Foreign-Born Migrants." *Population Today* 30, 6 (2002): 1, 4-5.

Gamboa, Erasmo. 1990. *Mexican Labor and World War II: Braceros in the Pacific Northwest, 1942-47*. Seattle/Austin: University of Washington Press/University of Texas Press, 1990.

Gamboa, Erasmo, and Carolyn Buan. *Nosotros: The Hispanic People of Oregon*. Portland: Oregon Council for Humanities, 1995.

Gamson, William, and Andre Modigliani. "Media Discourse and Public Opinion on Nuclear Power: A Constructionist Approach." *American Journal of Sociology* 95, 1 (1989): 1-37.

Ganz, Marshall. "Resources and Resourcefulness: Strategic Capacity in the Unionization of California Agriculture, 1959-1966." *American Journal of Sociology* 105, 4 (2000): 1003-62.

García, Juan Ramon. *Operation Wetback: The Mass Deportation of Mexican Undocumented Workers in 1954*. Westport, CT: Greenwood Press, 1980.

Garland, John. "The Oregon Forest Practice Act: 1972-1994." In *Forest Codes of Practice: Contributing to Environmentally Sound Forest Operations*. FAO Forestry Paper, 1996. FAO Corporate Document Repository. http://www.fao.org/docrep/w3646e/w3646e07.htm.

Geis, Sonya. "Forest Service Faulted for Lack of Outreach Programs." *Pasadena Star-News*, 5 December 2004.

Gozdziak, Elzbieta M., and Susan Forbes Martin. *Beyond the Gateway: Immigrants in a Changing America*. Lanham, MD: Lexington Books, 2005.

Gray, Gerald J. "Understanding Community-Based Forest Ecosystem Management: An Editorial Synthesis." *Journal of Sustainable Forestry* 12, 3 (2001): 1-23.

Greeley, William. *Forests and Men*. Garden City, NY: Doubleday, 1951.

Griffith, David. "Hay Trabajo: Poultry Processing, Rural Industrialization and the Latinization of Low-Wage Labor." In *Any Way You Cut It: Meat Processing and Small Town America*, edited by Donald D. Stull, Michael J. Broadway, and David Griffith, 129-51. Lawrence: University of Kansas Press, 1995.

Hamilton, Jim. "Feliz Navidad? Labor and Perspective in North Carolina's Christmas Tree Industry." Paper presented at the Annual Meeting of the Rural Sociological Society, Montreal, Canada, 27-29 July 2003.

Hansis, Richard. "The Harvesting of Special Forest Products by Latinos and Southeast Asians in the Pacific Northwest: Preliminary Observations." *Society and Natural Resources* 9 (1996): 611-15.

Hartzell, Hal. *Birth of a Cooperative: Hoedads, Inc*. Eugene: Hologos'I, 1987.

Harvey, David. *A Brief History of Neoliberalism*. Oxford: Oxford University Press, 2005.

Hayner, Norman S. "Taming the Lumberjack." *American Sociological Review* 10, 2 (1944): 217-25.

Haynes, Richard W., and Gloria E. Perez. *Northwest Forest Plan Research Synthesis*. Portland, OR: US Department of Agriculture, Forest Service, PNW Research Station, 2000.

Hernández-Léon, Rubén, and Victor Zúñiga. "'Making Carpet by the Mile': The Emergence of a Mexican Immigrant Community in an Industrial Region of the US Historic South." *Social Science Quarterly* 81, 1 (2000): 49-65.

Hill, M. "Adding Diversity to the Outdoors." *NY Newsday*, 25 August 2005.

Hirt, Paul. *A Conspiracy of Optimism: Management of National Forests since World War Two*. Lincoln: University of Nebraska Press, 1994.

"History of the American Tree Farm System." *Western Conservation Journal* 23, 2 (1966): 46-52.

Jenkins, Austin. "Rumors and Panic Follow Immigration Raids." *KUOW Program Archive*, 27 June 2007. http://www.kuow.org/program.php?id=13001.

Jenkins, J. Craig, and Charles Perrow. "Insurgency of the Powerless: Farm Worker Movements (1946-1972)." *American Sociological Review* 42, 2 (1977): 249-68.

Johnson, Kirk. "With Illegal Immigrants Fighting Wildfires, West Faces a Dilemma." *New York Times*, 28 May 2006.

Kalleberg, A. "Nonstandard Employment Relations: Part-Time, Temporary, and Contract Work." *Annual Review of Sociology* 26 (2000): 341-65.

Kalleberg, A., B. Reskin, and K. Hudson. "Bad Jobs in America: Standard and Nonstandard Employment Relations and Job Quality in the United States." *American Sociological Review* 65, 2 (2000): 256-78.

Kandel, William, and Emilio Parrado. "Hispanics in the American South and the Transformation of the Poultry Industry." In *Hispanic Spaces, Latino Places*, edited by Daniel Arreola, 255-76. Austin: University of Texas Press, 2004.

Kaplan, Amy. "Manifest Domesticity." *American Literature* 70, 3 (1998): 581-606.

Kemmis, Daniel. *Community and the Politics of Place.* Norman: University of Oklahoma Press, 1992.

Kesey, Ken. *Sometimes a Great Notion: A Novel.* New York: Viking Press, 1964.

Knudson, Tom, and Hector Amezcua. "The Pineros: Men of the Pines." *Sacramento Bee,* November 2005.

Koestler, Fred L. "Bracero Program." *Handbook of Texas Online.* Accessed 3 April 2006. Published by the Texas State Historical Association. http://www.tshaonline.org/handbook/online/articles/omb01.

Kusel, Jonathan, Susan Kocher, Jonathan London, Lita Buttolph, and Ervin Schuster. "Effects of Displacement and Outsourcing on Woods Workers and Their Families." *Society and Natural Resources* 13 (2000): 115-34.

"La Clinica fact sheet." Accessed 10 May 2011. http://www.laclinicahealth.org/News.asp.

Langston, Nancy. "Forest Dreams, Forest Nightmares: An Environmental History of a Forest Health Crisis." In *American Forests: Nature, Culture, Politics,* edited by Char Miller, 247-71. Lawrence: University of Kansas Press, 1997.

Lembcke, Jerry, and William Tattam. *One Union in Wood.* Madeira Park, BC/New York: Harbour Publishing/International Publishers, 1984.

Lemon, Sarah. "Nopales Grill." *Medford Mail Tribune,* 3 June 2005.

LeMonds, James. *Deadfall: Generations of Logging in the Pacific Northwest.* Missoula: Mountain Press Publishing, 2001.

Limerick, Patricia Nelson. *The Legacy of Conquest: The Unbroken Past of the American West.* New York: W.W. Norton, 1987.

Loose, Sarah. "The Workers behind the Wreaths." *Jefferson Center News* 4, 2 (2005): 1-3.

Lorensen, Ted. "The Forest Practices Act: Protecting Resources for 30 Years." *Forest Log* (July-August 2001): 14-15.

Lutz, Catherine A., and Jane L. Collins. *Reading National Geographic.* Chicago: University of Chicago Press, 1993.

Mackie, Gerald. "Success and Failure in an American Workers' Co-operative Movement." *Politics and Society* 22, 2 (1994): 215-35.

Mann, Geoff. "Race, Skill, and Section in Northern California." *Politics and Society* 30, 3 (2002): 465-96.

–. "The State, Race, and 'Wage Slavery' in the Forest Sector of the Pacific North-West United States." *Journal of Peasant Studies* 29, 1 (2001): 61-88.

Massey, Douglas S. *Beyond Smoke and Mirrors: Mexican Immigration in an Era of Economic Integration.* New York: Russell Sage Foundation, 2002.

Massey, Richard W. "A History of the Lumber Industry in Alabama and West Florida, 1880-1914." PhD diss., Vanderbilt University, 1960.

Maynard, Steve. "Rough Work and Rugged Men: The Social Construction of Masculinity in Working-Class History." *Labour/Le Travail* 23 (1989): 159-69.

McAdam, Doug. *Political Process and the Growth of Black Insurgency, 1930-1970.* Chicago: University of Chicago Press, 1982.

McAdam, Doug, John D. McCarthy, and Mayer Zald, eds. *Comparative Perspectives on Social Movements: Political Opportunities, Mobilizing Structures, and Cultural Framings.* Cambridge, UK: Cambridge University Press, 1996.

McArdle, Richard E., and Walter H. Meyer. *The Yield of Douglas-Fir in the Pacific Northwest.* Forest Service Technical Bulletin. Washington, DC: US Department of Agriculture Forest Service, 1930.

McCarthy, James. "Neoliberalim and the Politics of Alternatives: Community Forestry in British Columbia and the United States." *Annals of Association of American Geographers* 96, 1 (2006): 84-104.

McCarthy, John D., and Mayer Zald. "Resource Mobilization and Social Movements: A Partial Theory." *American Journal of Sociology* 82, 6 (1977): 1212-41.

McDaniel, Josh, and Vanessa Casanova. "Forest Management and the H-2B Guest Workers Program in the Southeastern United States: An Assessment of Contractors and Their Crews." *Journal of Forestry* 103, 3 (2005): 114-19.

McLain, Rebecca J., and Eric Jones. *Challenging "Community" Definitions in Sustainable Natural Resource Management: The Case of Wild Mushroom Harvesting in the USA*. London: International Institute for Environment and Development, 1997.

McWilliams, Carey. *Factories in the Field: The Story of Migratory Farm Labor in California*. Santa Barbara, CA: Peregrine Publishers, 1971.

Medford Mail Tribune. "Farmworkers Strike Threat." 1 July 1981.

Medford Mail Tribune. "Harvesting Hopes." 6 September 1981.

Medford Mail Tribune. "Labor Camps Rise in Northwest as Opposition Wanes." 7 July 1939.

Medford Mail Tribune. "Mexicans Will Work Here until First of Year." 9 November 1945.

Medford Mail Tribune. "Migrant Camp Is Solicited Here." 18 July 1941.

Menjívar, Cecilia. *Fragmented Ties: Salvadoran Immigrant Networks in America*. Berkeley: University of California Press, 2000.

–. "Impact of the Receiving Context: Salvadorans in San Francisco in the Early 1990s." *Social Problems* 44 (1997): 104-23.

Merchant, Carolyn. "Shades of Darkness: Race and Environmental History." *Environmental History* 8 (2003): 380-94.

Milkman, Ruth. *Organizing Immigrants: The Challenge for Unions in Contemporary California*. Ithaca, NY: ILR Press, 2000.

Millard, Ann V., and Jorge Chapa, eds. *Apple Pie and Enchiladas: Latino Newcomers in the Rural Midwest*. Austin: University of Texas Press, 2004.

Mitchell, Don. *The Lie of the Land: Migrant Workers and the California Landscape*. Minneapolis: University of Minneapolis Press, 1996.

Moseley, Cassandra. "Creating Community Benefit." EWP Briefing Paper Number 5. Institute for a Sustainable Environment, Ecosystem Workforce Program, University of Oregon, 2004.

–. "Ethnic Differences in Job Quality among Contract Forest Workers on Six National Forests." *Policy Sciences* 3, 2 (2006): 113-33.

–. *Procurement Contracting in the Affected Counties of the Northwest Forest Plan: Twelve Years of Change*. Portland, OR: US Department of Agriculture, Forest Service, PNW Research Station, 2006.

–. "Who Works in the Woods?" Paper presented at The Northwest Forest Plan: Ten Years Later, Portland, OR, 13 April 2004.

Moseley, Cassandra, and Stacey Shankle. "Who Gets the Work? National Forest Contracting in the Pacific Northwest." *Journal of Forestry* 99, 9 (2001): 32-37.

Murphy, Arthur D., Colleen B. Blanchard, and Jennifer A. Hill, eds. *Latino Workers in the Contemporary South*. Athens: University of Georgia Press, 2001.

National Employment Law Project. "California and Maryland Courts Uphold Undocumented Workers' Rights to Workers' Compensation." *Immigrants' Rights*

Update 19, 8 (2005). http://www.nilc.org/immsemplymnt/emprights/emprights094.htm.

New York Times. "Land Fraud in the West: Extensive 'Graft' Schemes Discovered on the Pacific Coast." 22 October 1903.

Noonan, Mark. "Raising Debate beyond Borders." *Risk & Insurance,* 3 March 2011. http://www.riskandinsurance.com/story.jsp?storyId=533333131.

Oregon Department of Consumer and Business Services. *Occupational Safety and Health in Oregon's Forests: Logging and Forestry Services.* Salem, OR: Oregon Department of Consumer and Business Services, December 2005.

Oregon Department of Forestry. "Figure 38-2. Estimated Annual Reforestation Spending 1982 Dollars." Accessed 16 May 2011. http://www.oregon.gov/ODF/STATE_FORESTS/FRP/crt6ind38.shtml.

Otis, Alison T., William D. Honey, Thomas C. Hogg, and Kimberly K. Lakin. *The Forest Service and the Civilian Conservation Corps: 1933-42.* Washington, DC: US Department of Agriculture, Forest Service, 1986. http://www.nps.gov/history/history/online_books/ccc/ccc/index.htm.

Pardo, Richard. "Community Forestry Comes of Age." *Journal of Forestry* 93, 11 (1995): 20-24.

Pawel, Miriam. "UFW: A Broken Contract." *Los Angeles Times,* 8 January 2006.

Peck, Jamie. *Work-Place: The Social Regulation of Labor Markets.* New York: Guilford Press, 1996.

Pellow, David. 2002. *Garbage Wars: The Struggle for Environmental Justice in Chicago.* Cambridge, MA: MIT Press.

Pfeffer, Max J. "Low-Wage Employment and Ghetto Poverty: A Comparison of African-American and Cambodian Day-Haul Farm Workers in Philadelphia." *Social Problems* 41, 1 (1994): 9-29.

Piore, Michael. *Birds of Passage: Migrant Labor and Industrial Societies.* Cambridge, UK: Cambridge University Press, 1979.

Piore, Michael, and Charles F. Sabel. *The Second Industrial Divide: Possibilities for Prosperity.* New York: Basic, 1984.

Portes, Alejandro, and Robert L. Bach. *Latin Journey: Cuban and Mexican Immigrants in the United States.* Berkeley: University of California Press, 1985.

Portes, Alejandro, and Rubén G. Rumbaut. *Immigrant America: A Portrait.* 3rd ed. Berkeley: University of California Press, 2006.

Preston, Julia. "Immigrants' Speedy Trials after Raid Become Issue." *New York Times,* 8 August 2008.

–. "Immigration Crackdown with Firings, Not Raids." *New York Times,* 29 September 2009.

–. "Iowa Rally Protests Raid and Conditions at Plant." *New York Times,* 28 July 2008.

Prudham, W. Scott. *Knock on Wood: Nature as Commodity in Douglas-Fir Country.* New York: Routledge, 2005.

Pulido, Laura. *Environmentalism and Economic Justice: Two Chicano Struggles in the Southwest.* Tucson: University of Arizona Press, 1996.

Rajala, Richard. "Bill and the Boss: Labor Protest, Technological Change, and the Transformation of the West Coast Logging Camp, 1890-1930." *Journal of Forest History* 33, 4 (1989): 168-79.

–. *Clearcutting the Pacific Rain Forest: Production, Science, and Regulation.* Vancouver: UBC Press, 1998.

–. "The Forest as Factory: Technological Change and Worker Control in the West Coast Logging Industry, 1880-1930." *Labour/Le Travail* 32 (1993): 73-104.

Ramakrishnan, S. Karthick, and Paul George Lewis. *Immigrants and Local Governance: The View from City Hall.* San Francisco: Public Policy Institute of California, 2005.

Reforestation Efforts in Western Oregon: Hearing before the Subcommittee on Forests of the Committee on Agriculture, House of Representatives, Ninety-Fifth Congress, First Session, July 8, 1977, Roseburg, Oreg. Washington, DC: Government Printing Office, 1977.

Robbins, William G. *Hard Times in Paradise: Coos Bay, Oregon, 1850-1986.* Seattle: University of Washington Press, 1988.

–. *Landscapes of Conflict: The Oregon Story, 1940-2000.* Weyerhaeuser Environmental Classics. Seattle: University of Washington Press, 2004.

–. "Lumber Production and Community Stability: A View from the Pacific Northwest." *Journal of Forest History* 31 (1987): 187-96.

–. *Lumberjacks and Legislators: Political Economy of the US Lumber Industry, 1890-1941.* College Station: Texas A&M University Press, 1982.

Robbins, William G., and Donald Wolf. *Landscape and the Intermontane Northwest: An Environmental History.* Portland, OR: Pacific Northwest Research Station, General Technical Report PNW-GTR-319, 1994.

Romm, Jeff. "The Coincidental Order of Environmental Justice." In *Justice and Natural Resources: Concepts, Strategies and Applications*, edited by Kathryn M. Mutz, Gary C. Bryner, and Douglas S. Kenney, 117-38. Washington, DC: Island Press, 2002.

Rothenberg, David. *With These Hands: The Hidden World of Migrant Farmworkers Today.* Berkeley: University of California Press, 2000.

Ruckelshaus, Catherine, and Bruce Goldstein. "From Orchards to the Internet: Confronting Contingent Worker Abuse." National Employment Law Project. 2002. http://www.nelp.org/content/content_publications/P600/.

–. *The Legal Landscape for Contingent Workers in the United States.* New York: National Employment Law Project, Farmworker Justice Fund, 2002.

Runyan, Jack L. *A Summary of Federal Laws and Regulations Affecting Agricultural Employers, 1992.* Washington, DC: US Department of Agriculture, Agriculture and Rural Economy Division, Economic Research Service, 1992.

Rural Voices Conservation Coalition. "Workforce and Labor Issue Paper." Sustainable Northwest. May 2006. http://www.sustainablenorthwest.org/resources/rvcc-issue-papers.

Salazar, Debra J., and Donald K. Alper. *Sustaining the Forests of the Pacific Coast.* Vancouver: UBC Press, 2001.

Sarathy, Brinda. "The Latinization of Forest Management Work in Southern Oregon: A Case from the Rogue Valley." *Journal of Forestry* 104, 7 (2006): 359-65.

–. "The Marginalization of Pineros in the Pacific Northwest." *Society and Natural Resources* 21, 8 (2008): 671-860.

–. "Minutes from Meeting between Matsutake Mushroom Harvesters and US Forest Service, Crescent Lake, OR." 30 July 2002.

Sarathy, Brinda, and Vanessa Casanova. "Guest Workers or Unauthorized Immigrants? The Case of Forest Workers in the United States." *Policy Sciences* 41, 2 (2008): 95-114.

Satterfield, Teresa. *Anatomy of a Conflict*. Vancouver: UBC Press, 2002.

Singer, Audrey, Susan Wiley Hardwick, and Caroline Brettell. *Twenty-First-Century Gateways: Immigrant Incorporation in Suburban America*. Washington, DC: Brookings Institution Press, 2008.

Smith, Scott T. "Word of Mouth: Las Morenas." *Medford Mail Tribune*, 20 January 2006.

Stauffer, J. "The President's Side." *ARC Quarterly* (Fall 1983): 6.

–. "The President's Side." *ARC Quarterly* (Fall/Winter 1984): 4.

Stephen, L. "Cultural Citizenship and Labor Rights for Oregon Farmworkers: The Case of Pineros y Campesinos Unidos del Noroeste (PCUN)." *Human Organization* 62, 1 (2003): 27-38.

–. *The Story of PCUN and the Farmworker Movement in Oregon*. In collaboration with PCUN staff members. Eugene: University of Oregon, Department of Anthropology, 2001.

–. *Transborder Lives: Indigenous Oaxacans in Mexico, California, and Oregon*. Durham, NC: Duke University Press, 2007.

Stull, Donald D., Michael J. Broadway, and Ken C. Erickson. "The Price of a Good Steak: Beef Packing and Its Consequences for Garden City, Kansas." In *Structuring Diversity: Ethnographic Perspectives on the New Immigration*, edited by Louise Lamphere, 35-64. Chicago: University of Chicago Press, 1992.

Swanson, Robert E. *Rhymes of a Western Logger*. Vancouver, BC: Lumberman Printing, 1943.

Swarts, Heidi. *Organizing Urban America*. Minneapolis: University of Minnesota Press, 2008.

Sweeney, Brendan. "Sixty Years on the Margin: The Evolution of Ontario's Tree Planting Industry and Labour Force, 1945-2007." *Labour/Le Travail* 63 (2009): 47-78.

Tejada-Flores, Rick, and Ray Telles. *The Fight in the Fields: Cesar Chavez and the Farmworkers' Struggle*. Directed by Rick Tejada-Flores and Ray Telles. Produced by Paradigm Productions. A presentation of the Independent Television Service. San Francisco, CA, 1997.

Terkel, Studs. *Hard Times*. New York: Pantheon Books, 1970.

Turner, Frederick Jackson. *The Frontier in American History* (New York: Henry Holt, 1921). Last modified 30 September 1997. http://xroads.virginia.edu/~HYPER/TURNER/.

Upson, Arthur. "Our Forest Resources Are Contributing to Victory." *Journal of Forestry* 40, 12 (1942): 909-13.

US Census Bureau. "The 2010 Statistical Abstract: Population." Last modified 27 October 2010. http://www.census.gov/compendia/statab/2010/cats/population.html.

US Department of Agriculture. "Migrant and Seasonal Agricultural Worker Protection Act." Last modified 30 September 2002. http://www.thecre.com/fedlaw/legal19/mspasumm.htm.

US Department of Labor. "Procedures for H-2B Temporary Labor Certification in Nonagricultural Occupations." General Administration Letter No. 01-95. 22 December 1997. http://www.ows.doleta.gov/dmstree/gal/gal95/gal_01-95c1.htm.

US Department of Labor. *Register of Wage Determinations under the Service Contract Act*. SCA No: 77-0079 Rev 33 Forestry and Land Management Services.

Washington, DC: US Department of Labor, Employment Standards Administration, Wage and Hour Division, 2004.

US Department of Labor, Bureau of Labor Statistics. "May 2005 State Occupational Employment and Wage Estimates." Last modified 24 May 2006. http://www.bls.gov/oes/2005/may/oessrcst.htm.

Vaughn, Jacqueline, and Hanna J. Corter. *George W. Bush's Healthy Forests: Reframing the Environmental Debate.* Boulder, CO: University Press of Colorado, 2005.

Voss, Kim, and Rick Fantasia. *Hard Work: Remaking the American Labor Movement.* Berkeley, CA: University of California Press, 2004.

Watson, Jack. "Minimum Wage." *ARC Quarterly* (Spring/Summer 1985): 5.

Weber, Edward. "A New Vanguard for the Environment: Grass-Roots Ecosystem Management as a New Environmental Movement." *Society and Natural Resources* 13 (2000): 237-59.

Welch, Craig. "A War in the Woods." *Seattle Times,* 6 June 2006.

Wells, Gail. *The Tillamook: A Created Forest Comes of Age.* Corvallis: Oregon State University Press, 1999.

Wells, Miriam. "Immigration and Unionization in the San Francisco Hotel Industry." In *Organizing Immigrants: The Challenge for Unions in Contemporary California,* edited by Ruth Milkman, 109-29. Ithaca, NY: ILR Press, 2000.

"What is A.R.C.?" *ARC Quarterly* (Spring 1982): 26.

Williams, Michael. *Americans and Their Forests: A Historical Geography.* Cambridge, UK: Cambridge University Press, 1989.

Wilma, David. "Weyerhaeuser Dedicates the Nation's First Tree Farm near Montesano on June 21, 1941." HistoryLink.org. Essay 5256. 21 February 2003. http://www.historylink.org.

Winslow, Carlile. "Wood and War." *Journal of Forestry* 40, 12 (1942): 920-22.

Winston, Steven. "A Letter to the Editor." *ARC Quarterly* (Fall 1983): 7.

Zybach, Bob. "Renewed Resources: The Reforestation of the Tillamook Burn." *ARC Quarterly* (Fall1983): 13-17.

–. "Safety and Profit." *ARC Quarterly* (Fall 1981): 6.

Index

The letter *t* following a page number denotes a table; *f* denotes a figure.

wage slaves, 11-12
wages: Bureau of Labor Statistics, 86, 157*n*35; farm workers, 61, 112-13; as marginality indicator, 82-83t, 86-89; overtime, 86; reforestation workforce, 29, 86; tree-planting, 37-38, 76, 88
Walport Ranger District, 43
War Production Board, 20, 21
Washington State: arrest of brush pickers, 98; forest management standards, 30; fruit circuit, 61; logging industry, 21, 23, 71; reforestation, 26-27, 29-30; timber industry, 15, 18, 80; worker retraining, 93
Weaver, Jim, 43
Weeks Act, 29
welfare capitalism, 79-80
West Coast Lumberman's Association, 148*n*38
wetbacks, 51, 53
Weyerhaeuser, Frederick, 17-18, 26-27
Weyerhaeuser Timber Company, 81
white pine, 16, 17
white pioneers, 6-7, 144*n*13
"white spaces," forests as, 6-8
wildland-urban interface, 73
Willamette Valley: background, 102-7; farm worker advocacy, 103; immigrant activist organizations,

105t, 110-11, 121-22, 121-27; Valley Migrant League (VML), 104
Willamette Valley Immigration Project (WVIP), 104, 107, 111, 115, 122
Wind River field station, 19
women, role of, 56-57, 61, 62, 66-67, 134, 137, 138
Woodburn, OR, 49, 122, 124
workers' compensation: history, 29; and Latino workers, 65, 76, 96; loggers, 79, 88, 89; Oregon, 151*n*80; tree-planting co-operatives, 36, 39, 40-41, 141
workplace injuries: Bureau of Labor Statistics, 84, 157*n*30; fatal/nonfatal injuries, 79, 81-86, 82-83t, 157*n*32; as marginality indicator, 82-83t, 85-86; reforestation workers, 76; reporting of, 85, 96; transportation-related fatalities, 84
World War II, timber production, 8, 20-21, 80
WVIP. *See* Willamette Valley Immigration Project

Yakima, WA, 49, 51, 76
Yolanda (Sister), 58
Young, Neil, 3
Youth Conservation Corps, 90

Printed and bound in Canada by Friesens

Set in Stone by Artegraphica Design Co. Ltd.

Copy editor: Joyce Hildebrand

Proofreader: Dianne Tiefensee

Indexer: Lillian Ashworth